SpringerBriefs in Molecular Medicine

For further volumes:
http://www.springer.com/series/13102

Christian Behl · Christine Ziegler

Cell Aging: Molecular Mechanisms and Implications for Disease

 Springer

Christian Behl
Christine Ziegler
Institute for Pathobiochemistry
University Medical Center of the
  Johannes Gutenberg University Mainz
Mainz
Germany

ISSN 2197-7925   ISSN 2197-7933   (electronic)
ISBN 978-3-642-45178-2   ISBN 978-3-642-45179-9   (eBook)
DOI 10.1007/978-3-642-45179-9
Springer Heidelberg New York Dordrecht London

Library of Congress Control Number: 2013955258

Printed on acid-free paper

Springer is part of Springer Science+Business Media (www.springer.com)

Aging is Unquestioningly Complex

(Thomas B. L. Kirkwood[1])

[1] Kirkwood TB (2011) Systems biology of ageing and longevity. Philos Trans R Soc Lond B Biol Sci 366(1561):64–70

Acknowledgments

First of all, the authors wish to think Michael Plenikowski for his excellent artwork.

Furthermore, gratitude is given to Christof Hiebel, Andreas Kern, and Bernd Moosmann for supplying data from their own work that are included and for many stimulating discussions on the topic.

The scientific work of the authors on aging which is cited and integrated in this book is supported by the Fritz und Hildegard Berg-Stiftung and the Peter Beate Heller-Stiftung, members of the Stifterverband für die Deutsche Wissenschaft.

Contents

Chapter 1
Aging and Cell Aging: An Introduction

Abstract Since more than 100 years people are constantly growing older and a further significant increase in life time is expected in the decades to come. A person born today has a high statistical chance to reach the age of 100, to become a centenarian. Since aging is the primary risk factor for many human disorders it is mandatory to understand the aging process and how it affects onset and course of disorders of the elderly. Scientifically the medium life span is discriminated from the maximum life span. While the latter is rather constant at approximately 120 years the medium life span is increasing. But not only the whole organism, also each single cell out of the billions making up our body has an individual life span ranging from days to months and years until it is eventually dying or exchanged. The majority of our nerve cells is never replaced. Understanding cellular aging and its influence on human disease is a key challenge of molecular medicine research.

Keywords Medium life span · Maximum life span · Life expectancy · Aging process · Age-associated disorders · Predisposition

The common understanding and the intrinsic meaning of the term "aging" is that it is a longer lasting process. This view, actually, applies to all organisms, when aging is considered as part of the life cycle of the individual species. Aging with very few exceptions occurs in all organisms and is characteristic for life. Some species live only for hours, some for over 100 years. The individual life span is specific for the respective species but this does not mean that all individuals of the same species have the identical life span. Interestingly, within species there is an inter-individual variability regarding the aging rate and the quality of aging which is a result from highly complex interactions of nature and nurture. "Nature and nurture" means the life long crosstalk between intrinsic genetic, extrinsic environmental as well as mathematically stochastic factors (Kirkwood 1999; Montesanto et al. 2012). There are many hypotheses, calculations, models and data collections that convincingly correlate the maximal age a species can reach with parameters such as body weight, body size, living conditions, metabolic rates etc. Many reviews and standpoints have been published on that very topic and still as always in science there are controversial

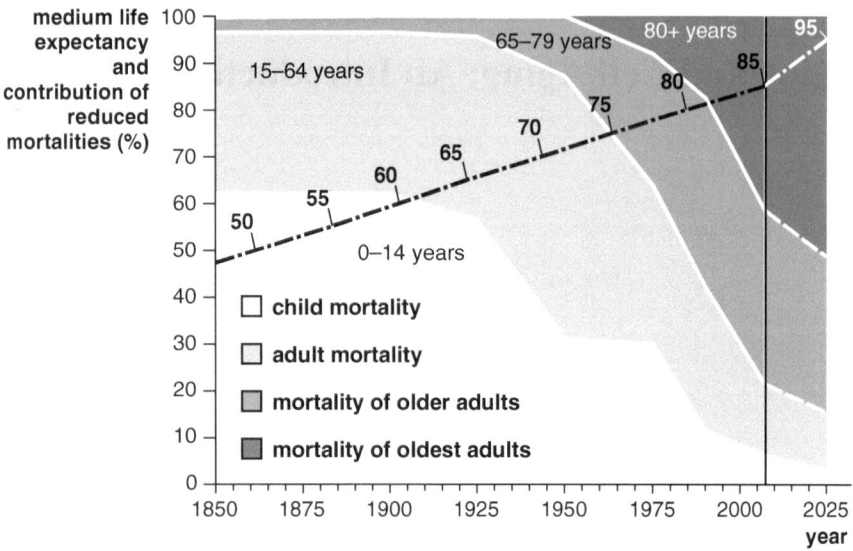

Fig. 1.1 Life expectancy over time. Medium life expectancy and the contribution of reduced mortalities from 1850 on; *dashed line* displaying medium life expectancy at birth (years) (women in selected industrialized countries; modified after Scully 2012)

arguments that are discussed and an ultimate final clue on aging has not been found yet and there is a good chance that it will never be.

When talking about life expectancy one has to differentiate between *medium life expectancy* and *maximum life expectancy* of a species. So, while the medium life expectancy of human beings in Western Europe has increased over the past centuries which is a result of, for instance, a reduced mortality during child- and adulthood and is currently approximately 82 years for females and 76 years for males (Fig. 1.1) the maximum life expectancy for humans is much higher. Madame Jeanne Louise Calment, the French lady who has been reported as the so far oldest human being confirmed, died at the age of 122 years and 164 days. In the meantime frequently one is facing reports on so called supercentenarians, and the number will increase. Up to date the age of Madame Calment is not reached by anyone else but very likely will be in the future. Obviously, the human organism can reach a maximum age of about 120 years. But still the medium life expectancy may change in the future, in particular due to a better understanding of the process of aging and, of course, the prevention and treatment of age-associated disorders (e.g. cancer) that may bring life to an early end or that may decrease the quality of the final years in life (e.g. Alzheimers Disease).

The medium life expectancy of the *Homo sapiens* has for many reasons dramatically changed over the last century and is still increasing (Fig. 1.1). The reasons for that are manifold and involve, for instance, a significantly better general hygiene early in life and reduced death rates of newborns, nutrition, the progress of medical care throughout life, and the possibilities of modern medicine (e.g. organ transplantation, tumor surgery, disease prevention, treatment of rare diseases, molecular diagnosis), meaning the improvement of general living and disease prevention and treatment conditions. As consequence there was a dramatic shift in the medium life span in humans, in particular along the last century where the quantum leaps especially in medicine occurred. The invention of penecillin by Alexander Fleming in 1928, for instance, was a giant step preventing untimely death upon bacterial infection. Modern medicine is driven forward by the developments of molecular biology, biochemistry and cell biology during the last 50 years or so. In fact, since the late 1950s and early 1960s, molecular biologists have uncovered many secrets of life and its control and have learned to structurally and functionally characterize, isolate, and manipulate the molecular components of cells and organisms. Today children learn already in elementary school that biology and biological processes are the result of the activities of molecules in our cells.

There is an obvious and intrinsic huge challenge resulting from the increasing medium life expectancy and the expectation to experience this longer life in good health. But even if one is not interested in further extending his own life and feels that a maximum of approximately 120 years might be enough for humans, when turning older one more intensively starts to consider the own family tree and the circumstances of death of close relatives, parents, grandparents, and grand-grandparents. So while it is fact that all human life ends, it is acknowledged that the personal life span is depending on family traits which basically means genetics. Also it is well accepted that the aging process is complex and influenced by many factors (aging is a multifactorial process) and that aging represents the ongoing interplay of an organism's biology with the environment in which the species live (concept of nature and nurture). The data derived from investigating individual family trees with methods of human genetics can give one a rather good guess of the personal life expectancy. A sophisticated analysis of specific target genes that are known to be linked to certain disorders that have eventually occurred in ones particular family tells a lot about disease risk during aging and may give suggestions for disease prevention. On top of this closer look on your own genes, one has to consider also additional secondary changes of the genome, epigenetic changes, that may occur already during fetal development before birth or later in life, for instance, as consequence of dramatic traumatic changes.

All together, the topic *aging* covers a complex array of scientific approaches and aspects ranging from genetics, molecular biology, medicine of aging to its philosophical meaning and the termination of human life. Depending on the scientific angle hundreds of different theories trying to explain aging were developed. There is no way to cover all views and angles how aging as physiological process but also as pathophysiological chain of events leading to disorders that are companions of increasing age is understood. In the context of this book aging is strictly considered as a physiological part of life. Various excellent reviews and books exist on aging looking at it from different sides (e.g. Kirkwood 1999; Finch 2007; López-Otín et al. 2013). Moreover, a topic of high significance that, ultimately, may then also mirror back to human aging, is the question of the fundamental differences of the life spans of different species ranging from hours and days to decades, also covered by excellent books and reviews that are discussed and referenced in above mentioned work.

The approach of this small book is different (1) by understanding aging of an organism (the human being) as aging of its molecules, its cells and organs, (2) by focusing on a selection of such topics that are related also to age-associated disease, and (3) by trying to find common nominators that can be assigned to different theories to get a clearer picture. Therefore, the very first part of this presentation is focusing on the aging process of the smallest unit of life, the cell itself. Next, this knowledge will be extended to the aging of tissue and the organism and connected to exisiting major aging theories. The presented mechanisms and views are discussed in the context of age-associated human disorders, mainly dealing with neurodegenerative disorders and cancer of the elderly. Again, molecular and cell biology have uncovered many different details on the aging process. Given that this small book cannot fully cover the tremendous amount of data that has accumulated over the last decades, nor follow all detailed and partially controversial discussions on certain topics, only some key mechanisms of cell and organism aging based on a personal selection can be discussed. And even when focusing "only" on the aging process of the cell it appears already that it is extremely complex and many signals and pathways are interdependently connected and intertwined and all of this in many cases is far from being fully understood.

1.1 Organ Systems Affected by Aging

Aging is the primary risk factor for a whole panel of human disorders (Fig. 1.2) and the understanding of the aging process will lead to better prevention, therapy and measures to increase resilience of virtually all main disorders associated with aging. What is known about the cause of major age-associated disorders? When screening through the literature of the last 20 years, for instance, with respect to Alzheimers disease (AD) it may appear that this still non treatable disorder has a well defined genetic basis. But as exciting and encouraging the identified early links (e.g. mutations in the gene for amyloid precursor protein; Brindle and St. George-Hyslop 2000) in some familial cases of AD may have been, the information does only apply to the strict genetic forms of the disease, which are approximately 1–3 % of all AD cases. But the majority (at least over 90 %) is of sporadic nature, does arise almost unexpectedly and is more likely to occur as we age. Recently, more and more molecular overlaps of the aging process and the pathogenesis of AD are discussed (Hunter 2013). It should be mentioned that the overwhelming set of genetic information on the familial cases of AD (Bettens et al. 2013) may have blinded research to some extent to focus also on a process that is admittedly hard to investigate such as aging. It is a well acknowledged fact that aging is the key risk factor for (sporadic) AD as we learn more about the cellular and molecular basis of aging. But only recently also the various molecular links between aging and

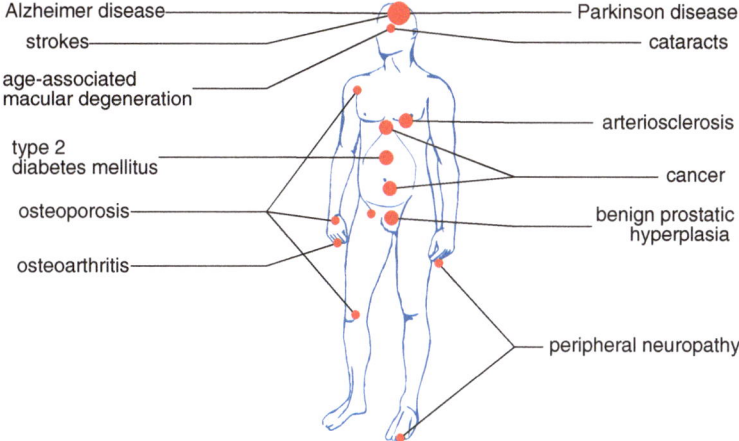

Fig. 1.2 Age-associated human disorders

AD-associated pathogenetic processes are combined also at the level of experimental research. As for AD such a change in research paradigms could also be applied to other strictly age-related disorders with so far unknown reason. The real cause(s) of AD may only be found when it is accepted that the aging biochemistry of a cell may have a great impact on the onset and progression of pathological process in later life. The focus on genetic linkage studies and in depth molecular analysis of genes and proteins derived from familial (genetic) cases will not solve the problem. It can be assumed that all of this applies also to other age-related disorders that display genetic and sporadic forms.

1.2 Age-Associated Functional Alterations

Human aging includes alterations in all organs leading to well known general patho-physiological phenotypes (Fig.1.3). Some of these disturbances of body functions are strictly age-associated. But minor impairment when further progressing can increase the chance of experiencing classically age-related deadly disorders, such as arteriosclerosis with stroke, heart failure, cancer and AD. For instance, short-term changes in lipid metabolisms can be counteracted and must not turn into a chronic pathological change, ultimately leading to vessel pathology and arteriosclerosis. But longer lasting changes may become risk factors for disease.

The biology of natural aging and the mechanisms of age-associated impairment and disorders are closely linked. And also a *successful aging* as demonstrated by the constantly increasing population of centenerians and supercentenarians finally leads to death that must have its cause in dysfunction of organs and, ultimately, of cells. Later in this discussion we will look into the biochemical changes that do

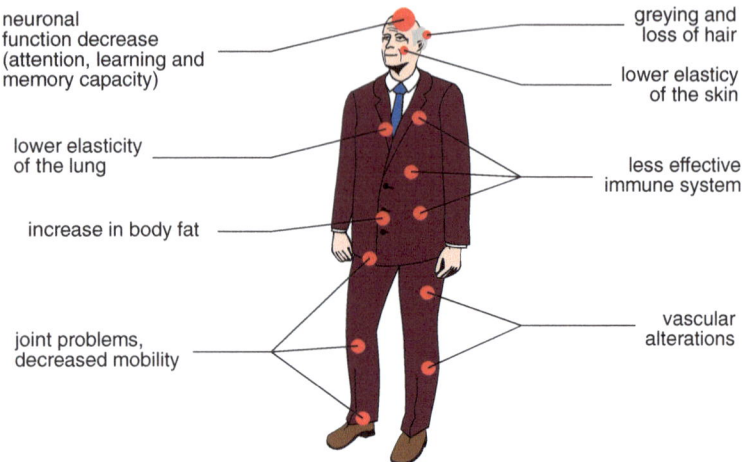

neuronal
function decrease
(attention, learning and
memory capacity)

greying and
loss of hair

lower elasticy
of the skin

lower elasticity
of the lung

less effective
immune system

increase in body fat

vascular
alterations

joint problems,
decreased mobility

Fig. 1.3   Age-associated alterations throughout the body

occur in cells and in cellular biomolecules that ultimately are well known causes for the functional changes during aging. One obvious and simple example of such biochemical changes occuring along with age at cellular biomolecules, for instance, is the oxidation of proteins that may lead to stiffening of joints and changes in general mobility (e.g. oxidation of so-called extracellular matrix proteins). Oxidation of biomolecules, proteins, lipids, and DNA is a key process that affects the cell as our life is depending on the particular chemistry of oxygen and organisms live under oxygen since its advent 2.3 billion years ago.

Age-related changes in our body may influence life quality and the predisposition to age-associated disease. One example: the dramatic decrease in estrogen levels in the female organism during and after menopause is such an age-related hormonal change with many different consequences (Gambrell 1982). Physiological estrogens (biochemically: estradiol, estrone and during pregnancy also estriol) are steroid molecules that mainly act via certain receptors, although the portfolio of possible interactions and the modes of activities of estrogens inside the cell is far wider than initially thought (Behl 2002; Faulds et al. 2012). Almost every cell of the body carries such estrogen receptors or is at least responsive in some way to estrogen. Therefore, estrogens affect many cellular functions. With the decline of the activity of the main estrogen producing organ in the female body, the ovaries, the concentration of estrogens in circulation and tissue is highly reduced. The strong reduction in this hormone then affects almost all organs and tissues in the female body, leading to different postmenopausal changes and unwanted effects in organ function (postmenopausal syndromes). It is well accepted that lack of estrogen is one factor driving the increased risk for e.g. osteoporosis after menopause. Given the medium life expectancy today females live approximately 40 % of their life under what can be called estrogen-reduced conditions. A loss of estrogen and estrogen-related functions may render organs more susceptible to other age-related changes and disease. Although the menopausal drop in estrogen is a natural and physiological process, it is a key environmental factor for the aging process of the female body and needs to be considered. In deed, low estrogen levels are discussed as a predisposition factor for a number of age-associated disorders (Christenson et al. 2012).

After introducing the terms *medium* and *maximum* life span above and giving some examples for age-related changes and a possible impact on pathology, the focus is now switched to the single cell in the human body. Interestingly, the life span of a cell can be days or can last throughout the whole life of the whole organism. Starting from so-called stem cell layers mucosa cells, for instance, face a life span of days, even shorter when facing exogenous insults. Erythrocytes, the red blood cells that are fully packed with hemoglobin a protein that carries oxygen throughout the blood stream from the lung to peripheral tissues, have a life span of approximately 120 days. Then they undergo a controlled cellular death program and are then renewed by the bone marrow, the stem cell tissue that gives rise to all blood cells. Different life spans in cells exist and the time range varies significantly: (1) the average life span of a typical liver cell, the hepatocyte responsible for manifold biochemical processes of the metabolisms (catabolism and anabolism), is around 5 months; (2) after their differentiation, neurons are present throughout the whole life of an organism. It is well known that the regeneration capacity of neuronal tissue, in terms of generating new cells, is rather limited and only a few highly defined areas in the nervous system have a real stem cell activity, for instance some selected areas in the mammalian hippocampus (Grandel and Brand 2013). One can easily imagine that the neurons forming the *nervus isciaticus* with extension sometimes longer than 1 m can not be exchanged during the human life span. The same applies to neurons in the neocortex and other brain areas where complex processes such as learning and memory, mood or fear are finding their cellular correlate.

Taken together our organism consists of billion of cells that are removed via selective mechanisms when their particular life span ends or upon damage, but there is also a significant number of cells that are never exchanged and stay alive and functional after their development and once they are completely differentiated. Every individual cell in our body is in a certain state and the individual aging of the different types of cells is depending on the general process called cell cycle that runs differently in erythrocytes, neurons, or tumor cells.

References

Behl C (2002) Oestrogen as a neuroprotective hormone. Nat Rev Neurosci 3(6):433–442
Bettens K, Sleegers K, Van Broeckhoven C (2013) Genetic insights in Alzheimer's disease. Lancet Neurol 12(1):92–104
Brindle N, George-Hyslop PS (2000) The genetics of Alzheimer's disease. Methods Mol Med 32:23–43
Christenson ES, Jiang X, Kagan R, Schnatz P (2012) Osteoporosis management in post-menopausal women. Minerva Ginecol 64(3):181–194

Faulds MH, Zhao C, Dahlman-Wright K, Gustafsson JÅ (2012) The diversity of sex steroid action: regulation of metabolism by estrogen signaling. J Endocrinol 212(1):3–12

Finch CE (2007) The biology of human longevity: inflammation, nutrition, and aging in the evolution of lifespans, 1st edn. Academic Press, Burlington

Gambrell RD Jr (1982) The menopause: benefits and risks of estrogen-progestogen replacement therapy. Fertil Steril 37(4):457–474

Grandel H, Brand M (2013) Comparative aspects of adult neural stem cell activity in vertebrates. Dev Genes Evol 223(1–2):131–147

Hunter S, Arendt T, Brayne C (2013) The senescence hypothesis of disease progression in alzheimer disease: an integrated matrix of disease pathways for FAD and SAD. Mol Neurobiol. 3 Apr 2013

Kirkwood TB (1999) Time of our lives: the science of human aging, 1st edn. Oxford University Press, New York

López-Otín C, Blasco MA, Partridge L, Serrano M, Kroemer G (2013) The hallmarks of aging. Cell 153(6):1194–1217

Montesanto A, Dato S, Bellizzi D, Rose G, Passarino G (2012) Epidemiological, genetic and epigenetic aspects of the research on healthy ageing and longevity. Immun Ageing 9(1):6

Scully T (2012) To the limit. Nature 492:S2

Chapter 2
Cell Cycle: The Life Cycle of a Cell

Abstract "Where a cell arises, there must be a previous cell". This early statement of Rudolf Virchow already points to the process that is called cell cycle. It describes a series of events leading to cell division and duplication and can be sectioned into phases that are controlled by a collection of proteins interacting with each other, the cyclines and the cycline-dependent kinases. It is mandatory that DNA replication is conservative meaning that its structure and sequence remain unaltered while the DNA is duplicated before the cell actually divides. Checkpoints are responsible for the supervision, proteins such as p53 and RB being the key protagonists in cell cycle control. Upon DNA damage recognized repair programs are activated or if repair fails the cell is driven into a programmed cell death to remove the damaged cell. The transfer of DNA mistakes from the mother cell to the daughter cells can lead to tumor formation. So p53 and RB are key tumor suppressor proteins.

Keywords Cell cycle · Cell cycle control · Cell cycle checkpoints · Cyclins · Cyclin-dependent kinases · Tumor suppressor proteins · Apoptosis · Cellular senescence

In 1858 Rudolf Virchow stated his famous cell doctrine "Where a cell arises, there must be a previous cell, just as animals can only arise from animals and plants from plants", meaning that cells are generated from cells, and only when the existing cells divide, more cells are arising (Virchow 1858). The cell cycle, also called cell-division cycle, describes a series of events that occur in a cell leading to its division and duplication and is, therefore, an identical replication. Eukaryotic cells, such as cells from mammalian organisms, do have a special organization structure, where the nucleus containing the genetic information (genome) can be distinguished from the surrounding cytoplasm containing different organelles and a huge assembly of proteins, the executors of the functions of the cell. During cell division an original cell ("mother cell") divides into two new cells ("daughter cells") in a highly controlled and organized manner (Fig. 2.1).

C. Behl and C. Ziegler, *Cell Aging: Molecular Mechanisms and Implications for Disease*,
SpringerBriefs in Molecular Medicine, DOI: 10.1007/978-3-642-45179-9_2,
© The Author(s) 2014

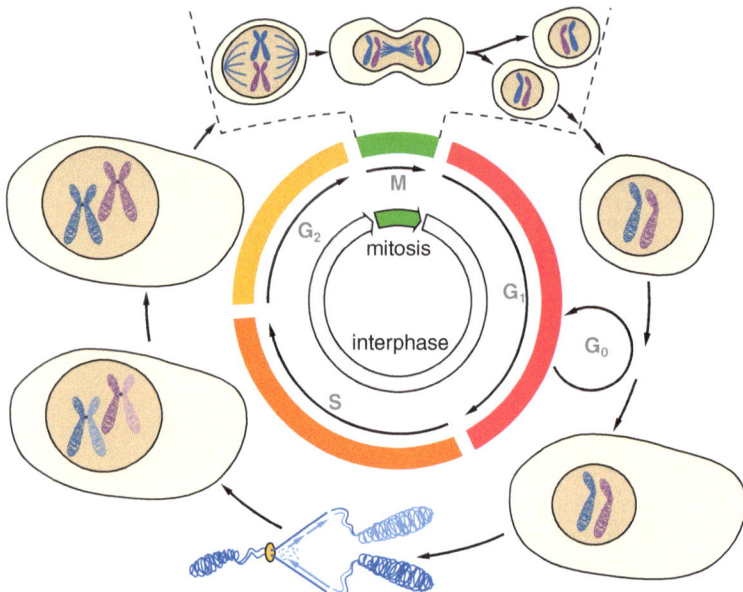

Fig. 2.1 Eukaryotic cell cycle. The division cycle of a cell can be sectioned into different pre-(S, G2) and postmitotic (G1, G0) interphases. Chromosome doubling and synthesis of other cell components are necessary before mitosis where the actual separation of a mother cell into two identical daughter cells occurs (M-phase). Cells that temporarily or reversibly escape the division cycle are in G0 phase (e.g. neurons; modified after Müller-Esterl 2011)

2.1 Phases of the Cell Cycle

Only cells with a nucleus undergo cycling and basically this eukaryotic cell cycle can be divided in two major phases, interphase and mitosis. While the interphase describes the time when the cell accumulates material and nutrients and subsequently doubles its genome, the mitophase summarizes the processes during which the cell actually splits into two distinct but identical daughter cells (identical DNA material, identical genome). The mitophase ends with the cytokinesis when the orginal (mother) cell is completely divided and the two daughter cells are on their own. Cycling of cells is a vital process for the whole body and the power of this event is best exemplified when looking at the first hours after a single-celled fertilized egg starts to develop, ultimately, into differentiated tissue and step by step into a full organism. But also many tissues in the adult body constantly go through cell cycling, a process that can be observed, for instance, when thinking of the renewal of hair, skin, blood cells and tissue of internal organs (e.g. hepatocytes in the liver). Cycling of the cell and ongoing or halted cell division is strongly related to cell and organism aging. It should be mentioned here that in studies on the life span of different species, the maximum life span of different animal models is put into relation with, for instance, body size, metabolisms, movement, and type of reproduction (sexual vs.

assexual). The hydra is a small freshwater polyp related to jellyfish that can achieve an indefinite life span if it reproduces asexually. Similar to yeast, this is possible via the budding off of daughter cells from the mothers body wall. So in these animals there is an unlimited high proliferative activity and cell cycling constantly going on (Deweerdt 2012).

Coming back to the biochemical process of the cell cycle. The major goal of this series of events is the doubling of the cell, for instance, for the replacement of damaged cells. Here, a major requirement is that the daughter cells are intact which relates mostly to the genomic material, the DNA. Since the mandatory process of DNA doubling (DNA replication) is a molecular process that underlies faults and errors as every biological process does, and because biomolecules are highly susceptible to chemical changes as, for instance, induced by UV light or radioactive radiation, DNA itself can be damaged during the process of cell cycling or errors occur during DNA replication. To maintain the correct function of the individual cell (e.g. skin cell, melanocyte) the DNA as the basis for all proteins must be maintained in its integrity, stability and function. This goal is reached via the execution of a tight control represented by DNA-repair enzymes etc., which are integral parts of the cell cycle. A look at the exact sequel of the steps of cell cycling will demonstrate the complexity of this process and will allow us to identify control mechanisms and checkpoints that may decide to either let the cell complete the cycle or let it run into death.

Many cells in our organism are permanently in cycle. Again, the division itself describes the process when one "mother cell" splits off into two genetically identical "daughter cells". Before this physical division, the so called phase of mitosis, the whole cell material needs to be duplicated. While the amount of protein material is increased by enhanced protein synthesis rates, the doubling of the cellular genome via DNA replication is complex and highly regulated. Roughly, there are two general major sequences in the cell cycle, the phases before and the phases after mitosis (M), called G-phases, where G stands for gap (Fig. 2.1). The gap phases are used to monitor the intracellular but also the extracellular conditions to make sure that everything is in order before actually proceeding to the next cycle phase. The G1-phase directly follows cell division and is frequently also called post-mitotic pre-synthesis phase. The cell starts to grow, the content of the cell (cytoplasm) with the functional machineries (organelles) is formed. In addition, the synthesis of mRNA takes place, histone proteins and the enzymes of the DNA replication machinery necessary for the next phase are generated. In a constantly dividing cell the G1-phase regularly takes approximately 3 h depending on the particular cell type. The S- or synthesis phase is characterized by the process of DNA replication (doubling of the cellular genome) and a major effort to produce histone proteins that are finally needed for the packaging of the genomic DNA. In average the S-phase takes about 7 h. The G2-, premitotic or post-synthetic phase is the time when the cell prepares to split off in two cells, the actual division. As part of a certain tissue cells loosen up the direct contact to neighboring cells, they usually round up and increase in general size. The synthesis of RNA and proteins concerns the main players needed for the mitosis and may take up to 4 h. Finally, in the M- or mitosis-phase division occurs, the doubled DNA organized in chromosomes is separated, the cellular nucleus divides (= karyokinesis) as does the rest of the cell (cytokinesis). The M-phase itself may take approximately 30–60 min and can itself be divided into five phases (pro-, prometa-, meta-, ana-, telophase; for details see Alberts et al. 2007). In proliferating tissue with cells undergoing constant divisions after mitosis then the next G1-phase occurs. Cells that are fully differentiated having special functions and roles in the tissue permanently remain in the G1-phase which is then called G0- or quiescence-phase. G0 represents a specialized resting and quiescent state of the cell. Nerve cells, muscle cells and red blood cells (erythrocytes) are the most prominent examples of cells in G0-phase. G0 is not only the state for differentiated cells so that they can fulfill their tasks in the tissue. Cells also enter G0 when the extracellular microenvironment is not in favor for further cycling, for instance when growth factors or nutrients are lacking that are necessary for the S-phase. In the context of this book it should be noted that a permanent arrest of cells in G0 is not identical with the senescence or the physiological aging of a cell and not all cells in G0 are on the road to death. Upon appropriate stimulation some cell types may re-enter the cell cycle. Damage to the cellular DNA and other significant changes provoke the quiescent state of the cell. Therefore, it is important to note that the status of senescence and of quiescence are different things, since once a cell is entering the senescence process this is a point of no return, ultimately leading to controlled cell death (apoptosis). Quiescence on the other hand is reversible. The state of a shorter or longer lasting quiescence is not only observed at the cellular level but also in whole organisms. One key model organism of aging research, the nematode worm *Ceanorhabditis elegans* (*C.elegans*) is developing in four larval phases. When the environmental conditions (e.g. low nutrition supply, too many further larvae) are not in favor of further proceeding in the development, the second larval state is changing into a permanent one, called *dauer*-state (dauer larvae) which can last up to 3 months. Interestingly, this quiescent state in *C.elegans* is induced by a specialized steroidal hormone (the so-called dauer-pheromone). The dauer-quiescence is actually a survival strategy of the worm development to overcome special or unfavourable conditions. While most nerve cells permanently remain in the G0-phase some cell types are in G0 for weeks and months and eventually reenter the cell cycle, including liver cells (hepatocytes) and lymphocytes. To do so cells need to be stimulated by special conditions and external growth factors. The knowledge about such external stimulatory signals that bring cells back into cycling or keep them permanently in the cell cycle is of pivotal importance for the understanding of various human disorders. Constant cycling triggered by uncontrolled growth input is a hallmark of cancer cells. As the understanding of any medical problem starts with the understanding of the basic molecular mechanisms and key control switches of pathological processes, here, the main executors of cell cycling should be shortly summarized.

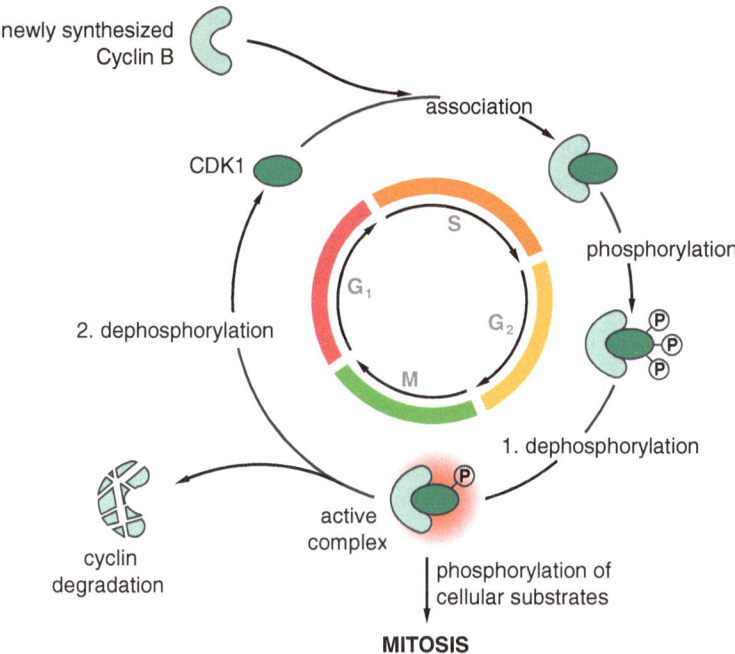

newly synthesized
Cyclin B

association

CDK1

S

phosphorylation

G_1

2. dephosphorylation

G_2

M

1. dephosphorylation

cyclin
degradation

active
complex

phosphorylation of
cellular substrates

MITOSIS

Fig. 2.2 Regulation of the cell cycle by cyclins/CDKs Cyclins and cyclin-dependent kinases (CDKs) interact physically. A sequence of phosporylation and dephosphorylation represents respective cell cycle phases. While after second dephosphorylation the CDKs enter a new round of cell cycling, cyclins become degraded and newly synthesized when needed (by the example of CyclinB/CDK1; modified after Müller-Esterl 2011)

2.2 About Cyclins and Cyclin-Dependent Kinases: Proteins that Trigger Cell Cycle Phases

The process of the cell cycle, the exact order of the phases as well as the metabolic and synthetic effort in the respective single phases needs to be strictly controlled. And cellular functions as well as their control and supervision are executed by specialized proteins. Two groups of proteins are essential for this cell cycle control, so-called cyclines and cycline-dependent kinases,CDKs (Fig. 2.2).

In more molecular detail: kinases are specialized enzymes that transfer phosphate groups mostly from ATP (adenosine triphosphate, generally speaking, the energy currency in cells) to specific substrates and are, therefore, phosphotransferases. While kinase is the name of the enzyme, the process itself is called phosphorylation and is one of the most frequent occurring post-translational modifications of proteins and essential for the regulation of the function of many proteins. More than 500 kinases are known in mammalian cells being key intracellular signal transmitters. In general, the residues of three amino acids in protein sequences are phosphorylated in most cases. Based on their biochemical structure these amino acids are threonine, serine and tyrosine. Kinase-driven phosphorylation can occur on multiple sites in proteins. And exactly this is happening in cell cycle control since a timely controlled transfer of phosphate groups and their removal which is biochemically performed by enzymes called phosphatases, actually controls the activity state of the cell cycle-associated proteins, the cyclines. Cyclines and CDKs are closely associated in the cell and the actual phosphorylation and dephosphorylation state of the cycline is the central regulation signal for the individual cell cycle phase. Today, at least eight types of cyclines (cycline A-H) and nine different CDK variants (CDK1-9) are described. Based on a huge amount of data, it is known that mainly cycline A-E and CDK1, -2, -4, -6 are directly affecting the cell cycle.

The cyclines can be divided into general classes that are each defined by the stage of the cell cycle in which they bind CDKs. The three classes that are essential in eukaryotic cells are: (1) G1/S-phase-cyclins that bind CDKs at the end of G1 and lead the cell to the replication of its DNA, (2) S-phase-cyclins binding CDKs in the S-phase and being essential for turning on the DNA replication, and (3) mitose-cyclins that promote and drive the process of mitosis. Since cell cycle control is excellently presented and discussed in major text books of molecular biology and only the basics can be addressed here, a short summary can be as follows: The actual cell cycle phase is determined by three major parameters, (1) the exact stoichiometric ratio of cyclines to CDKs, (2) the biochemical activity of these proteins determined by the phosphorylation state, and (3) the actual direct molecular (physical) interaction of these proteins (shown for cyclins in Fig. 2.2). The activity of the CDKs can be switched off by phosphorylations with inhibitory effects as well as by proteins, so called CDK inhibitor proteins (CKI, eg. p21 is a potent cyclin-dependent kinase inhibitor) that resemble the "controllers of the controllers" (for details, see Alberts et al. 2007). The control and monitoring systems of the cell cycle has been studied in molecular detail in yeast and the results are transferable to human cells. The cell cycle control proteins are highly conserved during evolution, so proteins of lower organisms and cells (e.g. yeast) can perform their function in mammalian cells and vice versa. So, we now know that the intimate cooperation combined with reversible phosphorylation of cyclins mediates and determines the actual cell cycle phase and, consequently, an upstream regulatory mechanism is necessary to control these cell cycle executors. Intracellular proteolysis is such a regulatory mechanism that controls the presence of proteins in cells. Most importantly, cyclin-CDK complexes are inactivated by regulated proteolysis that occurs via the ubiquitin-proteasome system, one of the two key protein degradation pathways in cells, which will be discussed also later in the context of their role for cell function and aging. The cell cycle-associated proteins are undergoing cyclical proteolysis to control their actual intracellular levels.

It was already mentioned that to maintain genome stability the DNA of the cycling and dividing cell needs to be transmitted without any changes or damage. Therefore, as soon as the cycle is started mistakes absolutely have to be avoided because otherwise changes in the genome would be transferred directly to the daughter cells. Consequently, all steps and phases of the cell cycle are closely monitored and several control and restriction points exist mediated by different classes of proteins. The main task of this control mechanisms is the prevention of the propagation of DNA misinformation that can lead to uncontrolled cell proliferation, ultimately causing tumor formation. To underline the key importance of cell cycle control mechanisms and its enduring significance for the understanding of pivotal processes in modern medicine and the development of disease the Nobel Prize in Physiology or Medicine has been given to Leland H. Hartwell, Tim Hunt and Paul M. Nurse in 2001 who all contributed to a better understanding of this fundamental process.

2.3 Better Save than Sorry: The Complex Control of the Cell Cycle

The control of cell cycling is a hot topic of molecular research although many details are already known. Here, the two main control proteins will be introduced, the retinoblastoma protein (Rb) and the protein 53 (p53). A lot of insight into cell cycle control has been gained by cancer research since, obviously, cells with unlimited proliferation as seen in cancer are the result of escaping from this control and any type of restrictions. Looking closer into cancer cells and their cell cycle it is found that frequently the control of G1 progression and S-phase initiation is disrupted, ultimately leading to an open and unrestrained entry into the cell cycle and proliferation. One important player in this context is a gene regulatory protein called E2F (a transcription factor) that binds to specific DNA sequences in promoters of genes that encode proteins necessary for the cell's entry into S-phase, including G1/S-cyclins and S-cyclins. Upstream its transcriptional activity the protein E2F itself is controlled by a direct interaction with a protein called Rb. Rb stands for retinoblastoma which are tumors of the human retina originally detected in an inherited form of eye cancer in children. Mutations in Rb lead to tumor formation, non mutated, i.e. wildtype Rb is suppressing tumor formation by interfering indirectly with the cell cycle (via E2F). Therefore, functionally intact Rb is a classical inhibitor of cell cycle progression. As shown in Fig. 2.3, during the G1-phase non-phosphorylated (active) Rb protein is associated with E2F. The Rb-E2F binding at the DNA inhibits the transcription of genes of the S-phase. In cells that are stimulated to proliferate, e.g. by extracellular growth factors, cyclin-CDK complexes accumulate leading to the phosphorylation and inactivation of Rb. The phosphorylation of Rb then causes a decrease in its affinity to E2F and finally the complete dissociation. E2F protein liberated from Rb leads to the activation of the transcription of S-phase genes. Taken together, by directly interacting with E2F and modulating its function, Rb is an inhibitor of cell-cycle progression preventing uncontrolled proliferation. Due to this prominent activity Rb is also called a tumor suppressor protein (Lombard et al. 2005). Considering that a single mutation in the gene coding for Rb leads to cancer development as seen in retinoblastoma underlines the key role for Rb as tumor suppressor. Rb does not stand for a single protein but rather for a family of proteins. But recently, the view of Rb as well known tumor suppressor controlling cell cycle progression and cell proliferation was extended. Interestingly, it has been found that Rb plays also a role

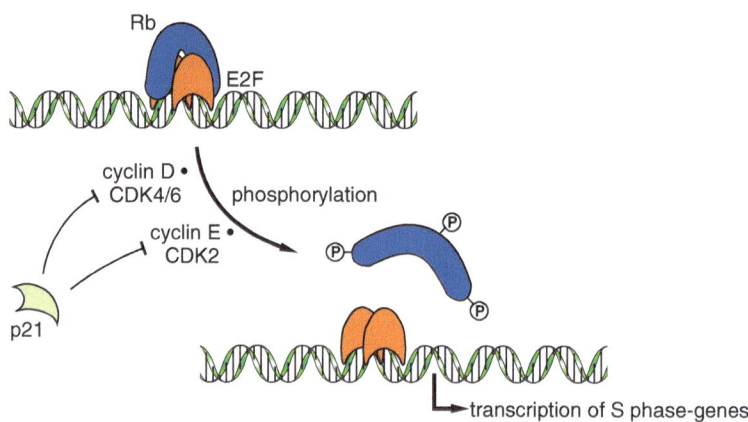

Fig. 2.3 The tumor suppressor Rb: Mode of action. The active retinoblastoma protein (Rb) binds to the transcription factor E2F. This Rb-E2F protein complex at the DNA blocks transcription of S-phase genes. Following stimulation Rb becomes phosphorylated (by the Cyclin D-CDK4/6 complex or CyclinE-CDK2 complex) leading to its displacement from E2F and the DNA. This dissociation allows transcription of S-phase genes (modified after Müller-Esterl 2011)

in the maintenance of genomic stability. Dysfunctional Rb protein drive the instability of chromosomes and aneuploidy, meaning an abnormal number of chromosomes (Manning and Dyson 2012).

Another key protein that prevents the uncontrolled proliferation of cells and, therefore, also acts as tumor suppressor, is the protein p53 (53 meaning its molecular weight of 53 kilodalton). Acting in a completely different way than Rb, p53 represents a checkpoint protein halting the cell cycle upon DNA damage. Cells are permanently confronted with a variety of external challenges. Of special interest is radiation, in particular, the ongoing UV and ionizing radiation caused by the atmosphere ("Höhenstrahlung") that both can cause direct damage to the DNA. Moreover, chemicals and toxins such as compounds that may intercalate into the double helix structure of the DNA may cause DNA damage, subsequently disturbing processes linked to the genomic DNA (e.g. transcription, replication). The most prominent and evident lesions at the DNA are crosslinks of the DNA helix double strand and other structural damage such as breaks of the DNA strand (discussed also later in the book). Beyond the disturbance of replication and transcription DNA damage can lead to changes in the DNA sequence, i.e. mutations that are if not removed conservatively inherited from the mother cell to the daughter cells which may again directly cause cellular deterioration, dysfunction and tumor formation. Since such DNA damaging events are rather frequent, cells do have intrinsic DNA repair mechanisms that can reverse damage and maintain the correct DNA structure and sequence. To allow this machinery which will be introduced later in some more detail the time for repair, upon DNA damage, cells in cycle are halted at the p53 checkpoint in the late G1-phase. p53 is a cellular key player that upon genotoxic and other stresses is activated by the upregulation of its protein level as well as by regulatory modulation (e.g. phosphorylation). p53 is involved in several pivotal signaling pathways, its specificity being governed by the interaction with other cellular proteins. Most important, p53 acts as a transcription factor inducing the expression of genes mediating growth arrest, DNA repair and apoptosis (see below). The G1 checkpoint is accomplished by p53 transactivating the CKI (cyclin-dependent kinase inhibitor) protein p21 that blocks G1/S-CDK complexes (Vousden and Lu 2002; Vogelstein et al. 2000; Tokino and Nakamura 2000). Moreover, it was shown that p53 takes direct part in the repair of double strand breaks by controlling the fidelity of recombination processes and thus exhibits functions counteracting carcinogenesis beyond cell cycle checkpoint control (Bertrand et al. 2004; Gatz and Wiesmüller 2006).

2.4 Last Exit Apoptosis

Mutations in the p53 gene leading to a dysfunctional gene product are very frequent in human tumors and observed in approximately half of all cancer pathologies. In case that the repair mechanisms are not sufficiently working or the DNA damage is just too severe, the intrinsic p53 control mechanism develops another function. Since during evolution the concept was successful that the health of the whole organism is more important than the survival of an individual cell which accumulates DNA damage and, subsequently, inherits mutated tumorigenic DNA putting the organism in danger, p53 activity can push such a cell into controlled cell death called apoptosis (Fig. 2.4). In fact, this particular function of p53 is a key for the prevention of cancer formation. It also demonstrates why in so many cancer types mutated and mal- or dysfunctional p53 is observed. Taken together, the correct function of p53 and Rb blocks uncontrolled cell division and actively prevents tumor formation. At the molecular level p53 and Rb are linked via the protein p21 (Figs. 2.3 and 2.4). The p53-mediated process to drive a damaged cell into the controlled suicide (apoptosis) can be seen as *exit strategy* to get rid of potential tumor precursor cells with accumulated DNA damage and to rescue the rest of the organism. In more recent literature it is also emphasized that "It is being recognized as a critical feature of mammalian cells to suppress tumorigenesis, acting alongside cell death programs" (Kuilman et al. 2010).

It is a key feature of tumor cells that they are running out of control and keep on passaging through the cell cycle and mitosis. But in most (non transformed) mammalian cells the potential to divide and, therefore, to run through cell cycles is limited indicating that there is a potential physiological endpoint of cellular life, the cellular senescence. This limited replicative potential of cells has been shown in primary cells put out of an organism into the culture dish and is called the "Hayflick limit" of mammalian cells (Hayflick and Moorhead 1961). In 1965 Leonard Hayflick further hypothesized that "the finite lifetime of diploid cell strains in vitro may be an expression of aging or senescence at the cellular level". Considering this view cellular senescence means a stable and long-term loss of proliferative capacity, despite continued cellular viability and metabolic activity (Hayflick 1965; for review: Kuilman

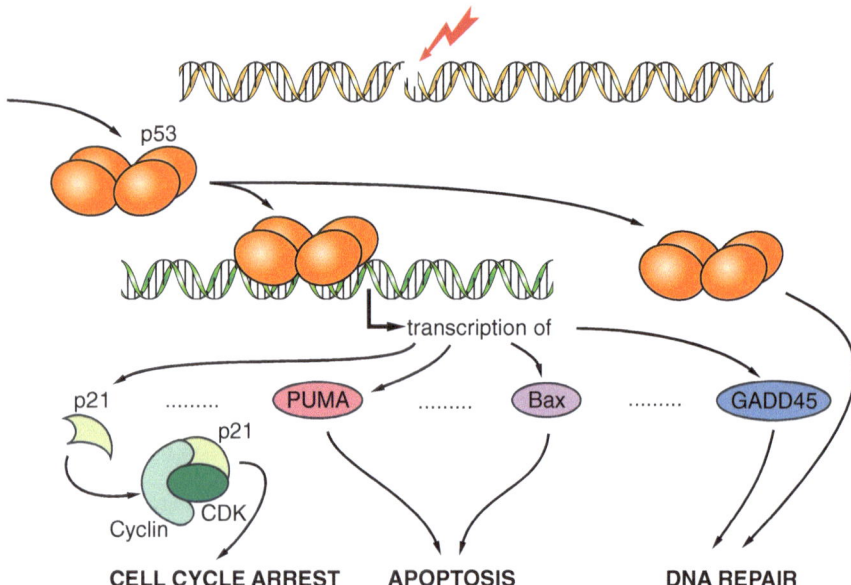

Fig. 2.4 The tumor suppressor p53: simplified mode of action. Upon DNA damage (e.g. strand breaks) as caused, for instance, by UV or ionizing radiation p53 is activated and induces the transcription of genes controlling cell cycle arrest (e.g. p21), apoptosis (e.g. Bax, PUMA) and DNA repair (e.g. GADD45). In addition p53 can also directly affect DNA repair

et al. 2010). Senescent cells can be identified also in tissues *in vivo*, where the actual cause why these cells become senescent is not exactly known. It is speculated that frequently senescence may be the result of the activities of activated oncogenes. The fate of senescent cells *in vivo* was unknown for long time, but recent new evidence suggests that these cells are cleared and removed by the innate immune system. Therefore, senescence and apoptosis can be also addressed as pathways occuring in parallel. Via both processes significantly damaged cells are eliminated from the body. Nevertheless, it is reported that some senescent cells persist in tissues and the number is increasing with age. It is currently hypothesized "that these persistent senescent cells have adverse effects on tissue function. If so, senescence may be an example of antagonistic pleiotropy, providing an anticancer mechanism in early life but having adverse effects on tissue function in late life" (Hornsby 2010).

In trying to explain the molecular basis of the clockwork of replicative senescence as consequence of the Hayflick limit one realizes that the limitation of the number of cell divisions may lie in the power and effectivenes of the enzymatic machineries of DNA replication and the particular physical structure of the chromosomes, namely their end structures, the telomeres. And this, actually is also the rational and basis of an important theory of aging, the *telomere theory of aging* that can be lined up as one key theory of aging which will be introduced in the next chapter.

References

Alberts B, Johnson A, Walter P, Lewis J, Raff M, Roberts K (5th edn) (2007) Molecular Biology of the Cell. Taylor & Francis, New York

Bertrand P, Saintigny Y, Lopez BS (2004) p53's double life: transactivation-independent repression of homologous recombination. Trends Genet 20:235–243

Deweerdt S (2012) Comparative biology: Looking for a master switch. Nature 492(7427):S10–1

Gatz SA, Wiesmüller L (2006) p53 in recombination and repair. Cell Death Differ 13:1003–1006

Hayflick L (1965) The limited in vitro lifetime of human diploid cell strains. Exp Cell Res 37:614–36

Hayflick L, Moorhead PS (1961) The serial cultivation of human diploid cell strains. Exp Cell Res 25:585–621

Hornsby PJ (2010) Senescence and life span. Pflugers Arch 459(2):291–299

Kuilman T, Michaloglou C, Mooi WJ, Peeper DS (2010) The essence of senescence. Genes Dev 24(22):2463–2479

Lombard DB, Chua KF, Mostoslavsky R, Franco S, Gostissa M, Alt FW (2005) DNA repair, genome stability, and aging. Cell 120(4):497–512

Manning AL, Dyson NJ (2012) RB: mitotic implications of a tumour suppressor. Nat Rev Cancer 12(3):220–226

Müller-Esterl W (2011) Biochemie: Eine Einführung für Mediziner und Naturwissenschaftler. Spektrum Akademischer Verlag, 2. Auflage

Tokino T, Nakamura Y (2000) The role of p53-target genes in human cancer. Crit Rev Oncol Hematol 33(1):1–6

Virchow R (1858) Die Cellularpathologie in ihrer Begründung auf physiologische und pathologische Gewebelehre. Hirschwald (Berlin), 1. Auflage

Vogelstein B, Lane D, Levine AJ (2000) Surfing the p53 network. Nature 408(6810):307–310

Vousden KH, Lu X (2002) Live or let die: the cell's response to p53. Nat Rev Cancer 2(8):594–604

Chapter 3
Theories and Mechanisms of Aging

Abstract The more one learns about single processes and genes known to be involved in aging, the more it becomes evident that these are connected and there is no unifying theory of aging. The individual theories put individual factors and processes in focus and for each theory there are direct links to life span or to age-related disorders. In the following chapter, the key theories of aging focusing on telomeres, DNA damage, oxidative stress as well as possible roles of nutrition, the interplay between genes and environment (epigenetics) and cellular protein homeostasis are presented. In animal models the life span can be altered by targeting specific genes, proteins and signalling pathways. After reviewing all these different mechanisms and factors obviously involved in the aging process of cells and organisms it becomes clear that aging is a multifactorial process where various intimate mutual interactions can be identified. Consequently, at the end of this chapter the idea of a molecular aging matrix composed of the major players affecting and triggering the aging process is developed.

Keywords Theories of aging · Telomeres · DNA damage · DNA repair · Sirtuins · Caloric restriction · Life span extension · Oxidative stress · Protein homeostasis · Epigenetics · Molecular aging matrix

If someone could explain, why exactly we are aging, this would be a significant progress in understanding the limitation of human life and likely would immediately call for ways to reverse it in tyring to reach unlimited life time. Such scenarios are in science fiction movies, at least, key plots but also for some scientific, pseudo-scientific and esoteric communities "anti-aging" is an important issue. The goal of such efforts is to counteract age-associated physiological changes of the body tissue and to extend the life span based on the individual wishes (see for instance http://www.antiaging.com; http://www.antiaging-systems.com). And the human phantasy has no boundaries when it comes to the key questions such as aging and end of life. This was nicely demonstrated in 2011 with the US-Movie *In Time*. The key plot here is that in the year 2169 all humans from birth on carry an intrinsic digital clock on their arm. At 25 years this very clock is activated and everyone has a residual life

C. Behl and C. Ziegler, *Cell Aging: Molecular Mechanisms and Implications for Disease*,   21
SpringerBriefs in Molecular Medicine, DOI: 10.1007/978-3-642-45179-9_3,
© The Author(s) 2014

span of exactly 1 year, meaning that for that very person the life then ends ("times out"). In this movie life time represents the only universal currency. One can gain life time by different measures (e.g. work), you pay with life time for products or services, as well as it can be transferred to others. The prerequisite of this movie plot is that every born human carries a special gene manipulation that was introduced before to avoid overpopulation on earth. Another example: the short story "The Curious Case of Benjamin Button" written by F. Scott Fitzgerald and first published in 1922 describes a person that is born with the appearance of a 70 year old man. When Benjamin reaches the age of 12, his family realizes that he is aging backwards turning younger and younger with the years. So even in this very old writing and also more recent science fiction life span and the aging process are addressed as being easily controllable and, therefore, manipulable on demand. We know today that this is not the case.

Many theories of aging, sometimes also synonymously called hypotheses of aging, exist; frequently in the aging literature including this presentation no epistemologic difference will be made concerning theory and hypothesis. To begin with, there is not a clear-cut and defined explanation for the aging process of the individual organisms. Already at the cellular level aging is highly complex. This applies then even more for the multicell organismal level. But there are biochemical processes and in some cases even single molecular players that obviously have a great influence on the course and speed of the aging process. They can promote or delay the aging process to some extent but they can not stop it. And, of course, there is no such thing like an aging switch that turns the aging process on or off. In the following paragraphs not all existing theories of aging can be introduced, there are just too many. But the focus will be on the (selected by the authors) most important mechanism and most influential factors of aging acknowledged in the scientific community. Meanwhile, it becomes increasingly clear and will also dominate the presentation here that basically all these key processes are interconnected at the molecular level and aging is of multifactorial origin. Hallmarks of aging and key biochemical processes are summarized that go along with aging and mechanisms that directly and indirectly affect life span of cells and organisms.

3.1 The Telomere Theory of Aging

The theory that takes the chromosomes and more specifically the telomeres of the chromosomes into focus is one of the early attempts to explain aging by concentrating on the basic unit of life, the single cell. Consequently, such cell-based theories imply that understanding aging of the cell may give important hints or may even explain aging of cell assemblies, tissues, organs and the whole organism. The so-called telomeres are the physical ends of the linear chromosome structure. Their specific nucleotide sequence and particular structure protect the chromosome endings against recombination via a side by side crossover which would actually lead to an exchange of DNA stretches and an overall change of the DNA sequence which needs to be stable and unaltered in dividing cells. Furthermore, the telomeres prevent the sticking, fusion and enzymatic degradation of the chromosome endings. During the process of mitosis telomeres are also involved in the recognition of the chromosomes and their separation in the meta- and anaphase of mitosis. Interestingly, external influences such as oxidations (as a consequence of "oxidative stress"; see below) can have significant impact on telomere integrity and function and shorter telomeres associated with enhanced oxidative stress were reported in type 1 as well as in type 2 diabetes (Vallabhaneni et al. 2013; Ma et al. 2013).

Telomeres get shorter with each round of DNA replication and cell division, ultimately explaining that cells enter the senescence state after a defined number of cell divisions (see: replicative senescence). It is simply the loss of DNA material at the chromosomal ending that is inducing senescence and cell death. Basically, cells have a great potential of recycling material and synthesing all biomolecules necessary for DNA replication and the division of a mother cell into two identical daughter cells. So why is the chromosome getting shorter with each cell doubling and what is the reason that the cell is not capable to maintain the chromosome's length and structures? The answer is that the enzymatic machinery that doubles the genomic DNA, the DNA polymerase, can not completely execute its function at the telomeres. The molecular reasons for this limitation in the enzymatic activity portfolio of the DNA polymerase lies in the site-directed mode of action of this enzyme (Cech 2004). Due to that, the DNA polymerase creates the so-called "DNA end replication problem" (Fig. 3.1). In consequence, with each replication round a short piece of the telomeres is lost leading to a constant cell division-dependent shortening of the chromosomes. When the chromosomal length is fallen below a critical length the intrinsic control mechanisms recognize this as damage to the DNA and prevent further divisions of the cell (Shay and Wright 2007), the cells life comes to an end. This means that the cell enters a senescence state without further divisions and finally transits to apoptosis (Harley and Sherwood 1997). "The ends of linear chromosomes pose a biological problem. During replication the lagging strand cannot be fully copied by standard polymerases". In the early seventies Olovnikov and Watson first pointed out the implications of this end replication problem. Again, by lacking a mechanism to replicate the ends of chromosomes, consequently, they will shorten with each cell doubling, eventually reaching a critical length leading to cell senescence and/or death of the cell (Olovnikov 1996; Corey 2009). But there must be exceptions to this telomere shortening concept since there are many particular cell types and tissues in our body that constantly divide because they need to do so, for instance germ cells or stem cells of regenerative tissue (e.g. oral mucosa stem cells). These cell types actually overcome the telomere shortening problem by expressing and activating a specific cellular enzyme, the telomerase, that compensates the DNA polymerase activity handicap at the end of the chromosomes and is specialized in the synthesis of the chromosomal end stretches. So this special enzyme, the telomerase, can solve the end replication problem and should be explained here shortly.

DNA replication, DNA polymerase, telomere shortening, telomerase: Genomic DNA of eukaryotes is generally double-stranded, meaning that it consists of two strands with in each case complementary nucleotide bases and running

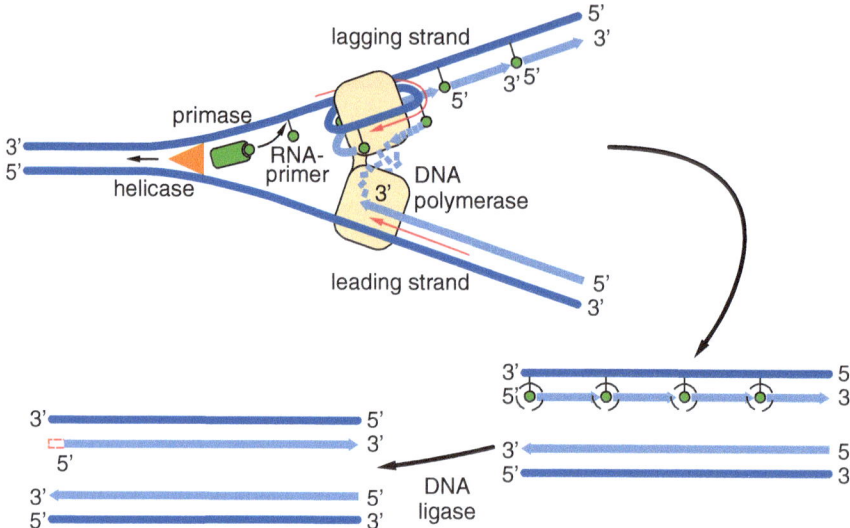

Fig. 3.1 The complex synthesis of DNA: leading strand, lagging strand and the end replication problem. By unwinding the DNA double strand by a helicase a replication fork is formed. While at the leading strand DNA is continuously synthesized by the DNA polymerase in $5' \rightarrow 3'$ direction, due to the opposite strand orientation doubling at the lagging strand is more complex and RNA primers are necessary for the step-by-step synthesis. In addition to DNA polymerase the enzyme DNA ligase is obligatory to finally create a complementary DNA strand . But on the very $5'$ end of this novel DNA strand a short piece of DNA is missing when the RNA primer is degraded. Consequently, DNA doubling at the end of the DNA strand and, therefore, at the end of the chromosome is not complete ("end replication problem")

in opposite directions in a biochemical sense ($3'$ to $5'$ and $5'$ to $3'$ respectively). The double-stranded DNA is replicated in a so-called seminconservative manner through a complex series of enzymatic reactions with the *DNA polymerase* δ being the key enzyme. Following the unwinding of the double strand via the enzyme helicase the polymeraseδ synthesizes the opposite strand but can do this only in one direction, from $5'$ to $3'$. So, while DNA synthesis is continuously running through at the "leading strand", at the "lagging strand" it consists of a set of reactions including RNA priming, synthesis of short DNA stretches, degradation of the RNA, filling in of DNA instead and ligation of the fragments. The ultimate consequence of this mode of DNA replication is always an overhanging $3'$ end of the DNA (see Fig. 3.1).

Whenever inside cells single-stranded DNA occurs it is a target of DNA degrading enzymes (DNAses). So when these enzymes cut off the single-stranded part of the replicated DNA, the newly synthesized and remaining $3'$-DNA end is obviously shortened. With each replication process the chromosome gets shorter for about 100 nucleotides. It is assumed that exactly this successive replication-dependent shortening of the chromosomal DNA is the molecular correlate for the limited division rate of cells, called the Hayflick limit. The chromosomes of the somatic cells of humans are equipped with a certain amount of telomere repeats (see below) but the telomerase activity is switched off in most cells. After many cell divisions the respective daughter receives significantly shortened chromosomes and is then removed from further cell cycling by the intracellular safeguards meaning the cell undergoes replicative cell senescence. The activity of the special enzyme telomerase solves the end-replication problem and it enables, for instance, stem cells, germ cells but also tumor cells to continously replicate their DNA and successfully and permanently run through and complete the cell cycle. Interestingly, bacteria are well-known to divide without any limitations as well (e.g. *E. coli* divides in the time window of approx. 20 min under ideal nutrient and temperature conditions) but do not face the end replication problem, since they carry circular DNA. It should be mentioned here that also the mammalian mitochondrial DNA consists of a circular molecule and does, therefore, also not have an end-replication problem.

The human telomere DNA consists of a stretch of several 1,000 nucleotides double stranded DNA with the repeating nucleotide sequence TTAGGG and a single-stranded stretch of 5–400 nucleotides overhanging at the so-called $3'$ end of the DNA (Alberts et al. 2007). Telomerase is a tricky enzyme complex containing a stretch of RNA (telomerase-RNA) that acts as a matrix to complete the TTAGGG repeat sequence at the telomeres. So, the enzyme brings its own template to bind to the $3'$-end of the DNA-strand and elongates it for exactly these 6 nucleotides, a process that is repeated several times. The rest of the work is then performed by other telomerase-independent enzymes that are present, including a DNA polymerase to complete the lagging strand resulting in a double-stranded chromosomal DNA end. This process is depicted in Fig. 3.2.

Taken together, the telomerase is a complex consisting of a protein enzyme and RNA. Its catalytic subunit is called hTERT which stands for human telomerase reverse transcriptase with the term reverse transcriptase referring to the ability of this enzyme to synthesize DNA from a RNA template. The activity of this enzyme complex is silenced in most somatic cells as mentioned above. But in most tumor cells telomerase is reactivated to ensure ongoing cell division. So, in addition afore cited mutations of the cell cycle safeguards p53 and Rb, the activation of the telomerase allows uncontrolled cell division and tumor formation. For human cancer cells it was shown that the expression of telomerase is indeed positively correlated with the aggressiveness of the tumors and their potential to metastasize. Today, it is acknowledged that one mode of action in the context of tumor formation is that hTERT may directly block apoptosis (programmed cell death), so that this physiological cell cycle escape pathway is prevented (Lamy et al. 2013). The discovery of the telomerase by Greider and Blackburn in the early 1980s caused great excitement (Greider and Blackburn 1985). Not only the molecular basis of one of the key questions in cell biology, namely how DNA replication and unlimited cell division can go on, was resolved but also the hope was fueled that with the targeted activation of telomerase cells could be pulled out of their fate facing replicative senescence and cellular aging could be stopped in general. Today, with the knowledge on a possible role of telomerase in tumor development and progression it is clear that the early expectations of anti-aging researchers were not fullfilled. Nevertheless, the description of

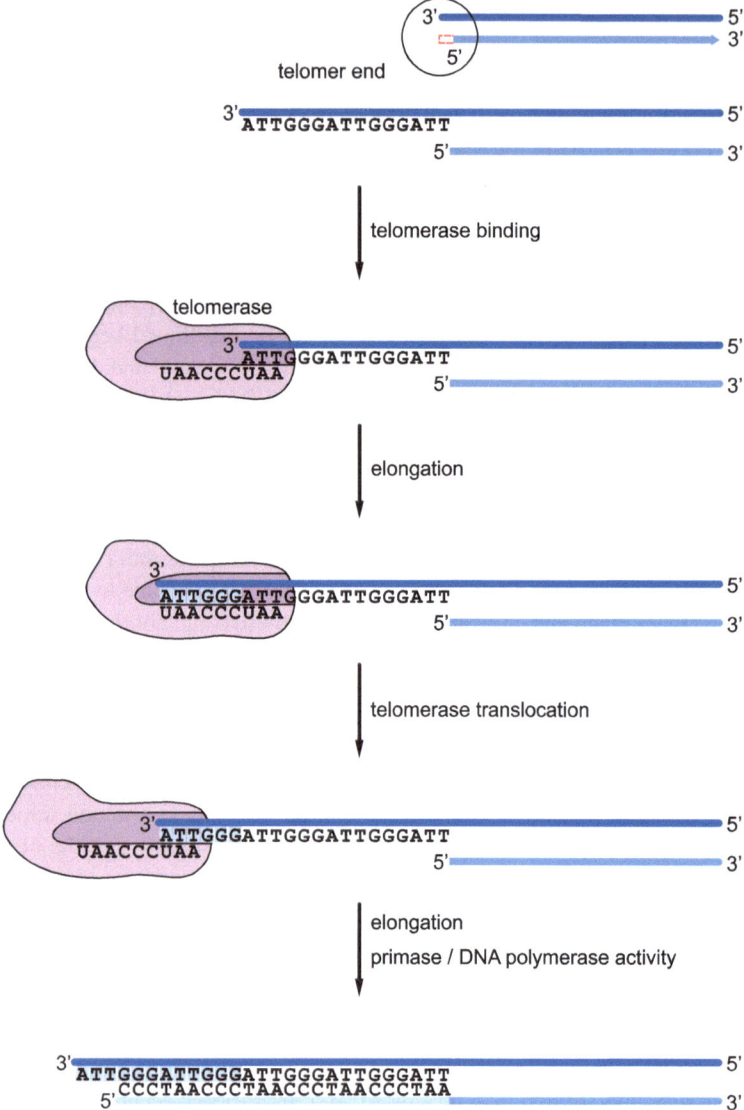

Fig. 3.2 Telomerase at work solving the DNA end replication problem. The enzyme telomerase consists of a protein and a short stretch of RNA that is complementary to the sequence at the overhanging 3′ end of telomeres. The telomerase activity is able to synthesize DNA from the RNA template whereupon the telomerase moves on, a step that is repeated several times. Then, the missing stretch can be filled in 5′ → 3′ direction (modified after Müller-Esterl 2011 and Alberts et al. 2007)

telomerase was of key importance for modern cell biology and, finally, Elisabeth H. Blackburn, Carol W. Greider and Jack W. Szostak received the Nobel Prize in Physiology or Medicine in 2009, for the discovery of "how chromosomes are protected by telomeres and the enzyme telomerase".

Telomeres, telomerase, aging, longevity and progeria syndromes: An interesting side aspect of telomeres in the context of aging is the direct comparison of the individual telomere length of different species. Mice and humans, for instance, have quite different telomer lengths. Telomeres of humans are approximately 10 kb in length which can be considered rather short when compared to mice with 20–50 kb long telomeres. In addition, while in humans during adulthood the telomerase expression and activity is low in almost all tissues, telomerase is present in most mouse tissues. This difference has been explained by the fact that the life span of mice is rather short meaning that the maintenance of an error-free genome is not necessarily as strictly controlled as it is for long-lived species as humans are. Before mice may develop tumors in their usual environment, the wilderness, they may end as prey for other animals. So tumor control systems may have developed as evolutionary advantage only in species that live long enough to be challenged by cancer development. Moreover, telomere length and telomerase activity have been correlated with body size and life spans (Seluanov et al. 2007).

The double-aged sword of modulating telomerase activity at an organismal level has been demonstrated nicely in transgenic mice. Upon an additional activation of telomerase by a transgenic over-expression of TERT an extension of the life span of approximately 10 % has been demonstrated (in a group of mice). On the other hand this in vivo manipulation and increase of telomerase activity is leading to a higher incidence of tumors. These tumors lead to an enhanced mortality in the first years of life and, obviously, the 10 % life-span increase refer to surviving mice (González-Suárez et al. 2005). Only an additional overexpression of tumor suppressor genes (e.g. p53) leading to more cancer-resistant mice reduces the early-life mortality and increases life span (Tomás-Loba et al. 2008). So, experimentally a clear link between telomerase and aging is shown. But also in humans this has been described. In a population of Ashkenazi centenarians, a special haplotype of telomerase is present leading to longer telomeres and an overall longer life (Atzmon et al. 2010). Interestingly, a number of human syndromes with enhanced or progressive aging, so called progeria syndroms or *progeria*, is known that exhibit an enhanced telomere shortening. Such pathological changes of normal aging in humans may serve as human aging models and studying progeria surely may give direct hints to the mechanisms of aging. Another interesting aspect when looking at progeria is that depending on the specific type different molecular changes cause these syndromes. So progerias have different causes but share the aging phenotype strongly suggesting that multiple causative factors and molecular changes are involved in human aging.

Prominent examples of progeria are the Werner-Syndrom, the Hutchinson-Gilford-Progeria-Syndrom and others (Klapper et al. 2001). Both clinical conditions display very early in life different age-associated phenotypic characteristics ranging from pathological changes such as arteriosclerosis, diabetes and dementia to less dramatic but still typical age features such as early greying and loss of hair and wrinkle formation of the skin. Werner-Syndrome patients at an age of, for instance, 40 years look already as being 70 or 80 years. The molecular cause of the Werner-Syndrom is a well-defined genetic defect in an enzyme involved in DNA replication, the DNA helicase (also called Werner-helicase). The disease clinically starts between age 10 and 20, when the typical pathological changes start to develop such as short stature, cateracts, hair greying and loss, scleroderma-like skin alterations, later on accompanied by osteoporosis, arteriosclerosis, neoplasms and type-I-diabetes. Werner syndrome patients have a median life span of approximately 50 years. The mutations in the Werner-helicase are so-called loss-of-function mutations but, interestingly, Werner-helicase may also have a function in the repair of the telomeres. For the Werner-Syndrom, the Bloom-Syndrom and Hutchinson-Gilford-Progeria-Syndrom (HGPS) the enhanced shortening of the telomeres is believed to be causally involved in the pathology.

The progression of HGPS is much faster and the life span that is reached by HGPS patients ranges from only 7 to 27 years of age. The genetic cause of HPGS was found to be a mutation in the gene coding for the protein lamin A/C. Lamin A is part of the nuclear envelope where it plays a key role in shaping the nucleus within cells. Lamin A mutations that cause HGPS lead to the production of an abnormal version of the lamin A protein resulting in an unstable nuclear envelope. Ultimately, these structural changes progressively damage the nucleus leading to premature cell death. Today a number of different such mutations is known and the resulting disorders are summarized as laminopathies. A closer look into the pathological changes that do occur in HGPS reveals also links to other age-related cellular changes. Today it is known that the mutation in the lamin A gene is leading to an altered mRNA transcript and, consequently, to an aberrant lamin A protein, also called progerin (Moulson et al. 2007). Intact lamin A is present at different sites in the nucleus and may there influence different processes. It is not only crucial for general nuclear integrity, it influences also transcription, may affect epigenetic modifications (epigenetics, see later in this book) and DNA replication. In addition it is also speculated that mutant lamin A has a direct effect on telomere length (Decker et al. 2009). As consequence of the different roles of lamin A in the nuclear compartment where various essential key processes are constantly going on different mechanisms of the pathogenesis of HGPS and certain links to biochemical processes involved in normal aging are currently discussed. These include also (1) altered telomere dynamics, (2) enhanced DNA damage (e.g. by oxidative stress) and defective repair, and (3) altered cell proliferation and senescence (Chavez et al. 2009; Burtner and Kennedy 2010).

Another subset of progeria syndromes is directly linked to a deficient DNA repair (Alberts et al. 2007). Among those is the rare but nevertheless mechanistically interesting progeria syndrom Ataxia Telongiectasia (AT). This disease is caused by a defective x-ray-responsive DNA damage-sensing protein kinase that activates p53 via phosphorylation (ATM kinase). The previously presented p53 checkpoint in the cell cycle thereby is lost. Cells with defective DNA escape the safeguard system and are not removed leading to cancer development. In addition, immune deficiency and brain degeneration is observed already in young age (McKinnon 2012). As it can be seen in AT DNA damage that is manifested instead of being repaired can cause progeria. When looking even further at the complex world of DNA repair enzymes and repair mechanisms it appears that there are several other inherited syndromes with defects in DNA repair. Because the individual cell invests a great effort to maintain the intact structure and sequence of its DNA, with regard to the DNA damage theory of aging it is easily conceivable that mutations in genes coding for DNA repair enzymes can also directly lead to progeria and increased cancer development. In progeria an accelerated accumulation of damage to DNA is characteristic (Lombard et al. 2005; Burtner and Kennedy 2010; Freitas and de Magalhães 2011). A big portion of genomes (several percent in the genomes of bacteria and yeast) are known to code for proteins involved in DNA repair. Due to its assumed major role in aging and disease next the DNA damage theory of aging will be presented.

3.2 The DNA Damage Theory of Aging

Genomic DNA and the propagation of the genetic information to the next generation of cells and organisms are key to life. It is well acknowledged that during aging of mammalian cells DNA mutations and DNA damage accumulate. As mentioned above the majority of human syndromes associated with accelerated aging (progeria) are caused by mutations of genes involved in the functional maintenance and integrity of the nuclear DNA and in DNA repair.

The *DNA damage theory of aging* is rather old but its core assumptions are still very prominent and have strong scientific foundations. In simple words, this theory proposes that the key to the functional changes associated with aging is DNA damage accumulation over (life) time leading to a misbalance in cellular homeostasis (Szilard 1959). In this early paper it is stated: "Our theory assumes that the elementary step in the process of aging is an 'aging hit', which 'destroys' a chromosome of the somatic cell, in the sense that it renders all genes carried by that chromosome inactive. The 'hit' need not destroy the chromosome in a physical sense. We assume that the 'aging hits' are random events and that the probability that a chromosome of a somatic cell suffers such a 'hit' per unit time remains constant throughout life. We further assume that the rate at which chromosomes of a somatic cell suffer such 'hits' is a characteristic of the species and does not vary appreciably from individual to individual. As a result of an aging process of this nature, the number of the somatic cells of an individual organism which have 'survived' up to a given age (in the sense of remaining able to fulfill their function in the organism) decreases with age. On the basis of our assumptions... the 'surviving' fraction of the somatic cells decrease with age at an accelerating rate" (Szilard 1959). What is called 'hit' in this early paper may refer to the randomly occurring DNA damage as induced by UV light or ionizing radiation outlined earlier. Later on the assumption that DNA damage as an event that is distinct from a specific mutation of the DNA is the primary cause of aging was further elaborated on by Alexander (Alexander 1967) as will be outlined below.

In fact, all cellular biomolecules experience direct or indirect damage leading to structural and functional changes. Oxidative stress, the burden of the life of cells and organisms in an oxygen atmosphere, can cause structural changes in biomolecules due to the high chemical reactivity of oxygen free radicals that can directly oxidize lipids, proteins as well as DNA. But in contrast to lipids and proteins no general replacement or constant turnover of the DNA is possible. Very briefly, it can be summarized: (1) Proteins in their correct three-dimensional structure serve complex functions for quite some time but they do have a certain half-life. When chemically modified proteins can lose their conformation and function. A frequently occurring chemical modification is the oxidation of amino acid side chains in proteins leading to structural changes. But, the cell performs powerful protein degradation processes that will be pointed out later in more detail. (2) Lipids that have been oxidized can be recycled via enzymatic systems and the cellular metabolism is almost constantly involved in the synthesis and degradation of fatty acids, the latter for the generation of energy. So with respect to proteins and lipids, in the case of chemical modifications that lead to structural changes and dysfunction there is an active and ongoing cellular lipid turnover depending on the metabolic demand of the individual cell. In general, the so-called metabolom describing the set of metabolically active biomolecules (lipids, peptides, proteins) is highly flexible and shows a great amount of plasticity. (3) The cellular DNA on the other hand, the nuclear DNA carrying the information of the genome, is present and ideally unaltered throughout the life cycle of the cell and, therefore, has to cope with and defend itself against chemical modifications as induced by acute and chronic oxidative stress or radiation. As presented in the discussion of the cell cycle, when DNA damage is too extensive, cells are removed from further cycling and the apoptosis program. If transcription of genes and their function is impaired this may halt the cell cycle and lead to cell death or, if the cell retains DNA damage, to a loss of gene expression and disturbed intracellular homeostasis. But not all damage is long-lasting since effective DNA repair mechanisms are active as part of the key effort of the cell to maintain the integrity of the human genome.

Of course, alterations of the DNA as they are introduced by mutations are crucial for the evolutionary process. Without any changes in the genomic DNA modified protein function is occurring that represent changes that can be of evolutionary advantage leading to a positive (or in the reverse case a negative) selection of the whole organism. Mutations in the DNA consist of DNA base deletions, base insertions, base exchange and general base pair rearrangements, all leading to the generation of an altered mRNA by transcription and an altered protein by the translation process. So, DNA mutations affect the information fixed in the DNA. In contrast to DNA mutations DNA damage describes a physically altered and/or chemically modified DNA structure. Although related DNA damage and DNA mutations are distinctly different: DNA damage is frequently the basis for mutations as it can cause errors during DNA synthesis. Here, known types of DNA damage will be shortly summarized. Before doing so it should be mentioned that the focus here is on DNA damage (and repair) of the DNA of the nucleus. Nevertheless, damage to the circular DNA of the cellular mitochondria is also discussed as important regulator of aging. And, indeed, compared to the genomic DNA in the nucleus the mitochondrial DNA is potentially much more challenged by oxidations since the mitochondrial respiratory chain frequently generates free oxygen radicals. In addition, it is known that mitochondria can not rely on the wide portfolio of DNA repair mechanisms that are found in the nucleus (see below) and, consequently, there is a high mutation frequency in the mitochondrial DNA (Cline 2012; Gredilla et al. 2012). Moreover, the packaging of mitochondrial DNA is quite different compared to the nuclear DNA which is actually based on the interaction of nuclear histone proteins that directly interact with the DNA. On the other hand, the mitochondrial DNA encodes only the small number of 37 genes indicating the limitations of the potential impact of damage to the mitochondrial DNA as compared to the genomic DNA in the nucleus.

Types of damage to the nuclear DNA: Interestingly, in mammalian cells DNA damage is not a rare event but occurs rather frequently. As estimated for mice more than 1,000 DNA lesions happen every hour in every cell (Vilenchik and Knudson 2000), a number that is also recognized for human cells. As stated by pioneers of this research field "DNA damage appears to be ubiquitous in the biological world, as judged by the variety of organisms which have evolved DNA-repair systems" (Gensler and Bernstein 1981). The sources of DNA damage can be of exogenous as well as of endogenous origin. Among the external agents that cause DNA damage the most important are UV light, ionizing radiation (cosmic, gamma, X-rays), mutagenic chemicals and viral infections. Intrinsic damage can be due to spontaneous chemical reactions but mainly is driven by metabolic byproducts such as mitochondria-derived reactive oxygen species. All the agents cause typical types of damage, partly also several of these at a time. Due to the complex structure of the DNA and its constituents, the purine and pyrimidine bases and the phospodiester back bone, DNA damage is polymorphic. The most frequent types resulting from cellular processes are oxidation, alkylation (usually methylation), deamination and loss (depurination, depyrimidination) of bases. Among exogenously caused DNA damage the formation of cytosine and thymine dimers by UV-B light and double strand breaks provoked by ionizing radiation are the most important types. Single strand breaks that are dominant in number are caused by various agents. Environmental chemicals can generate diverse types of DNA damage. A highly relevant example for an extrinsic chemical agent that damages the cellular DNA is cigarette smoke which is actually a complex mixture of chemicals with many genotoxic (DNA damaging) lung carcinogens (cancer causing). The most prominent mode of action of cigarette smoke-based carcinogens is the induction of DNA adducts (Hecht 2012).

The consequences of structural changes to the DNA are obvious, including problems with replication, transcription and the manifestation of mutations. The non-enzymatic methylation of the nitrogen-containing DNA bases generates nucleotides that are frequently base-pairing not correctly. For instance O-6-methylguanin pairs rather with thymine instead of its natural pairing partner cytosine resulting in a GC→AT transition when the DNA is replicated (Fig. 3.3).

Repair of the nuclear DNA: There is a saying that quite nicely describes the potential and significance of DNA repair mechanisms: "mutation is rare because of repair" (source of citation unknown). Mutations are changes in the DNA sequence

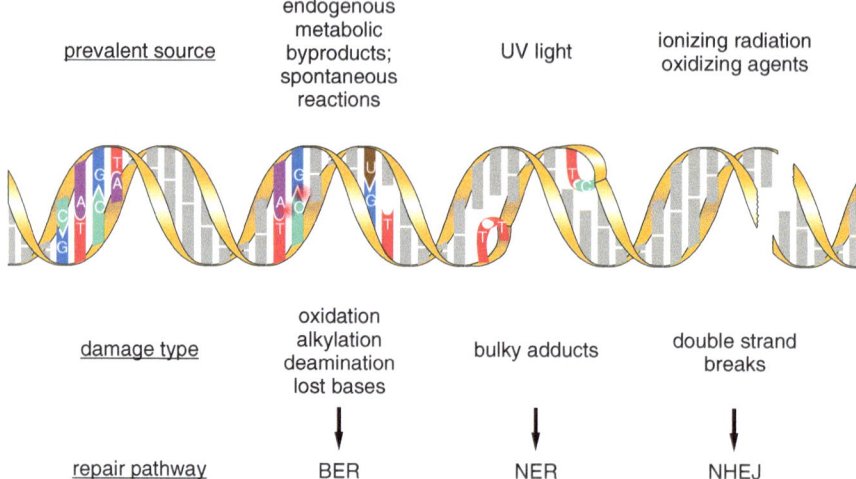

Fig. 3.3 Selected causes and types of DNA damage. Different endogenous and exogenous sources constantly introduce different types of DNA damage that can be counteracted by certain DNA repair pathways (*BER* base excision repair, *NER* nucleotide excision repair, *NHEJ* non homologous end joining; modified after Hoeijmakers 2001)

(often as a result of DNA damage) that successfully escaped the cell cycle control points and the DNA repair mechanisms. Different mechanisms and pathways of how a cell can repair a recognized damage to its DNA are known. Here, the focus is only on those three major types of DNA repair processes that have also described links to the aging process: base excision repair, BER, nucleotide excision repair, NER, and double-strand break repair by the process of non-homologous end joining, NHEJ. For a detailed overview of all so far known types of DNA repair pathways including also mismatch repair, single-strand break repair, homologous recombination repair, the reader is referred to the textbook or expert reviews on this topic (Alberts et al. 2007; Branzei and Foiani 2008; Masui and Kuramitsu 2010; Freitas and de Magalhães 2011; Kim and Wilson 2012; Curtin 2012). In addition to the relevance of defects of DNA repair processes in cancer development (Curtin 2012) more and more a role for a disturbed DNA repair is discussed also for neurodegenerative disorders (Cleaver et al. 2009; Jeppesen et al. 2011).

There are two forms of BER: short-patch BER with the replacement of only one single nucleotide and the long-patch BER with the excision and replacement of 2–13 nucleotides (Krokan and Bjørås 2013). It is suggested that this type of DNA repair is of particular importance for the nervous system and the brain and it has been implicated in the pathology of age-associated neurodegenerative disorders (Sykora et al. 2013). The metabolism of the brain is highly active and strictly depending on oxygen consumption and, consequently, oxidative stress is prominent. Nervous tissue is highly vulnerable to oxidations and BER is the main pathway for the repair of DNA damage caused by oxidative stress. BER is a repair pathway that remains

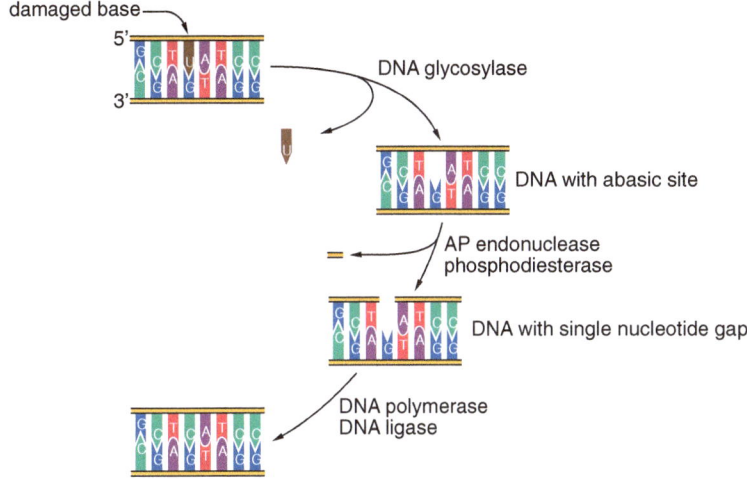

Fig. 3.4 Base excision repair, BER. In BER the enzyme DNA glycosylase removes only the damaged base, a subsequent series of enzymatic activities replacing it. Shown here for desaminated cytosine (C) which is concordant with uracil (U) normally used in RNA and when present in DNA leading to mispairing (modified after Alberts et al. 2007)

stable and functional in post-mitotic neurons. With respect to links to aging the evidence for an involvement of BER is not too strong and also conflicting to some extent. While a general reduction of the activity of BER with aging was observed in different tissues (Xu et al. 2008) there are also reports finding a higher expression of BER mediating enzymes during aging (Lu et al. 2004). In mice that are deficient in the gene sirtuin 6 and therefore show strongly accelerated aging, genomic instability and defects in BER activity have been observed (Mostoslavsky et al. 2006). Sirtuin 6 is an enzyme (actually a histone deacetylase and mono-ADP ribosyltransferase, see below) responsive to stress and affects pathways involved in aging, namely including DNA repair and the maintenance of telomeres, as well as in cellular metabolism and inflammation (Beauharnois et al. 2013) (Fig. 3.4).

There are also two forms of NER, the global genome NER and the transcription-coupled NER. In fact, the mechanisms of recognition of the DNA damage are different and two distinct groups of proteins are involved in that part of the process in each subtype pathway. Following DNA damage recognition, the two subpathways converge for the next steps including dual incision, repair, and finally DNA ligation. NER can recognize several types of DNA lesions caused by external stimuli. Whereas the global genome NER (as indicated by its name) is active all over the genome, the transcription-coupled NER occurs in genes that are transcriptionally active (Kamileri et al. 2012; Curtin 2012). Rather strong evidence for a link between NER and aging comes from studies showing that defects in different enzymes active in NER cause prominent progeria in humans, namely *Xeroderma pigmentosum*, *Cockayne syndrome* and *Trichothiodystrophy*. In addition, many mutations in genes linked to this type of DNA repair result in strongly accelerated aging in mouse models (e.g. Niedernhofer 2008; Lehmann et al. 2011). Focusing on UV light-induced DNA damage models the age-dependent NER activity was investigated with opposing results showing an age-associated decrease as well as an age-associated increase in NER activity (Freitas and de Magalhães 2011) (Fig. 3.5).

The third DNA repair mechanism with described links to aging is non-homologous DNA end-joining (NHEJ) which is, actually, the central pathway of the cell to repair DNA double-strand breaks. NEHJ is active throughout the whole cell cycle (Lieber et al. 2003; Lieber 2010). The name non-homologous DNA end-joining is owed to the fact that the DNA strand break ends are directly ligated by the NEHJ enzymes. There is no requirement for a template as this is the case in the process explicitly called homologous recombination repair depending on the usage of the intact sister chromatid (Moore and Haber 1996). NHEJ is a process that can be divided basically in three basic steps. Shortly, the end binding and tethering involves the binding of the so-called KU protein heterodimer to the broken DNA strand ends. In the next step, the *end processing*, damaged or mismatched nucleotides are removed by nucleases and DNA is synthesized by the enzyme DNA polymerase. Finally, DNA ligase is active and carries out the ligation of the DNA (Fig. 3.6).

The NHEJ DNA repair process is evolutionarily well conserved. NHEJ usually is imprecise which consequently leads to gene diversification. The observed incomplete accurateness is suggested to also contribute genetic changes leading to cancer development and aging (Lieber et al. 2003; Lieber 2010). A decrease in the level of KU protein complexes has been described in Alzheimers disease as well as normal human aging (Kanungo 2013). In addition several mouse knock out models with a deficiency in different KU genes indicate an effect of the KU complex on aging (Freitas and de Magalhães 2011). Double-strand breaks are accumulating in neurons of aged rats and there also the activity of NEHJ is decreasing (Vyjayanti and Rao 2006; Rao 2007).

It is of obvious interest to study the efficiency of such DNA repair and stress defense systems in the constantly increasing population of very old (over 100 years of age) people, the so called centenarians. In fact, Chevanne and colleagues compared the repair of DNA strand breaks in lymphocytes derived from young people, old individuals and centenarians by challenging cultured cells from the different groups with oxidative stress. The study showed that repair in cells (lymphocytes) from centenarians is as effective as in those derived from young individuals. Differences in protein expression levels were observed in the samples from the centenarians including an increased amount of the KU protein (KU70) (Chevanne et al. 2007). In a similar study DNA breaks were analyzed in lymphocytes from subjects of different age groups (20–35, 63–70, 75–82 years) and likewise the resistance of lymphocyte DNA to oxidative stress-induced damage and the resulting repair activity were measured. The outcome was quite interesting: the investigators found "an increase in oxidative base damage in old age, but this apparently does not result from deterioration of either antioxidant defence or DNA repair. In fact, both of these tend to increase with age". These results suggest the possibility that in old age DNA repair and also the antioxidant defense systems are induced in consequence (as compensation) of an increasing

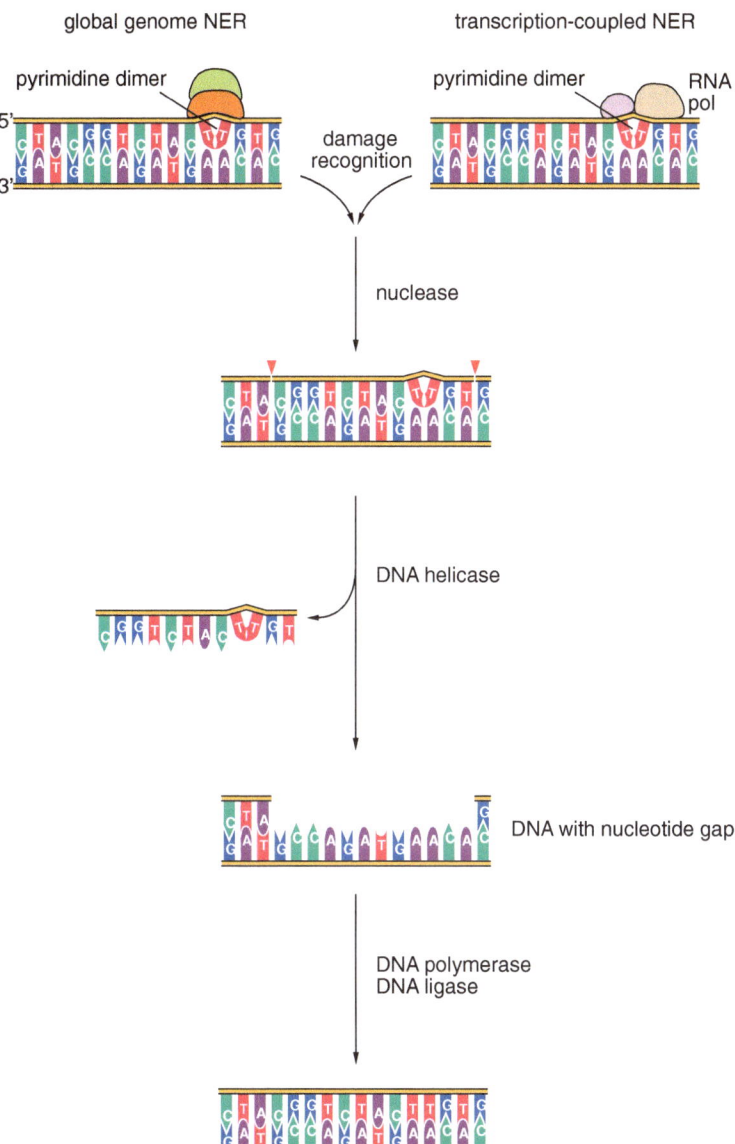

Fig. 3.5 Nucleotide excision repair, NER. Two forms of NER are known, the global genome NER and the NER that is coupled to the transcriptional process. After damage recognition nucleases remove a whole stretch of the damaged DNA strand which is in contrast to BER where single bases are removed. The resulting DNA gap is then filled by the activity of DNA polymerase and DNA ligase (modified after Alberts et al. 2007)

Fig. 3.6 Non homologous end joining, NHEJ. Double strand breaks to DNA are repaired with the help of Ku proteins and additional factors that transiently "seal" the broken single strands and introduce enzymatic activities (nuclease, polymerase, ligase) that remove the bases at the site of fracture and reform a DNA double strand. These processes are mechanistically flexible and lead to distinct sequence results, one option shown here (modified after Lieber 2010 and *MIGL* A database dedicated to understanding the Mechanisms of Intron Gain and Loss, University of Pittsburgh 2012; http://cpath.him.pitt.edu/intron/intronlossGenomicDeletion.html, Sept-22 2013)

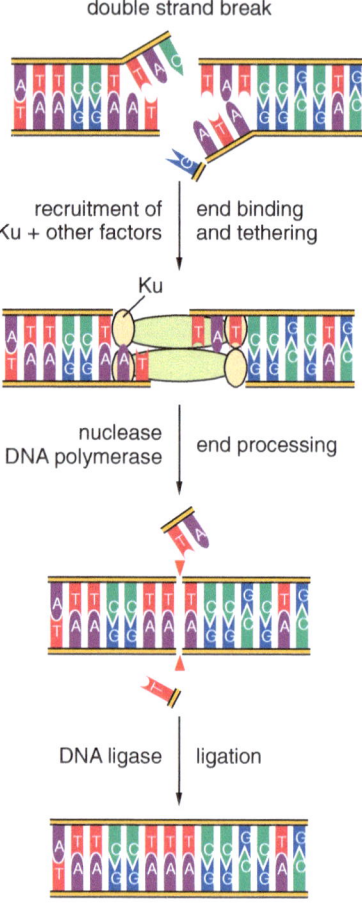

age-associated challenge. Alternatively, as the authors state "it is equally possible that older people, as survivors, had relatively high levels of antioxidant defenses and DNA repair earlier in their lives, compared with those who did not survive to such an age" (Humphreys et al. 2007). So the "survivors" may have adapted to the increasing challenge by the upregulation of defense and repair systems. Although these results are highly interesting, the data are so far only correlative and it is very important to study also the genetics, metabolism and the basic cell biology of cells from centenarians. In search of the genetic basis of a healthy long life many different studies on gene associations and single nucleotide polymorphisms (SNPs) are performed. As recently demonstrated for age-associated disorders genome-wide association studies resulted in a wide array of genes that are possible disease susceptibility factors (Jeck et al. 2012). But when looking at the different genes that can be associated with longevity in humans it seems again that many different factors and conditions can potentially contribute. In a recent meta-analysis evaluating different genetic studies in the search for longevity and healthy aging genes the authors also state: "The genetic contribution to longevity and human aging is likely to result from many genes each with modest effects. Some genes will likely affect longevity by increasing susceptibility to age-related disease and early death, whereas other genes are likely to slow the aging process itself leading to a long life. How genetic factors and their interaction with modifiable behavioral and environmental factors contribute to longevity remains unknown" (Murabito et al. 2012).

In the discussion of DNA repair mechanisms and the regulation of this machinery it is necessary to underline that the actual rate and efficiency of DNA repair depends on the cell status (mitotic or post-mitotic), on cell age, as well as on external and internal (metabolic) factors. And despite effective repair some DNA damage can remain and accumulate, a fact that is more pronounced in post-mitotic cells such as neurons and heart muscle cells. As we learned, in general, DNA lesions are rather frequent than rare events and the repair systems are efficient but eventually not efficient enough. But what happens to a cell that eventually accumulates a certain amount of damage? In fact these cells can face three immediate potential scenarios: (1) they are sorted out of the cell cycle and enter programmed cell death (apoptosis) mediated by certain cell cycle control proteins, (2) they can, indeed, escape the damage control systems and run into uncontrolled cell division and, ultimately, tumor formation, (3) the cells can enter the state of senescence, thereby being withdrawn from the cell-division cycle but remaining metabolically active and, ultimately, switch into apoptosis later or are cleared by components of the innate immune system.

To summarize again, the DNA damage theory of aging states that aging is caused by naturally occurring DNA damage and lesions that are not repaired and, consequently, accumulate over time. Damage to the nuclear DNA can contribute to aging either indirectly via increasing apoptosis or pushing cells into senescence or directly by increasing cell and organ dysfunction (Fig. 3.7). Given that DNA lesions are rather frequent events the DNA damage theory of aging is very conclusive and the idea that the accumulation of DNA damage over (life) time may drive the aging process makes full sense. On the other hand, as it is pointed out also in this book, there are many other mechanisms and key players that can induce aging in cells and model organisms that are completely independent of DNA lesions.

Of course, the evolution of the species is driven by small genetic variations but after a mutation has succeeded then again genetic stability is important. The process of DNA replication alone is highly complex, needs many players that are under a fine-tuned control. As we have learned, most of the accidentally occurring DNA lesions caused also by, for instance, metabolic changes, irradiation or environmental toxins, are not seen because they are instantly repaired by a set of enzymes belonging to the DNA repair systems (Alberts et al. 2007; Germann et al. 2012; Casorelli et al. 2012; Jena 2012; Yi and He 2013). Moreover, one thing that could be learned when looking closer into progeria syndromes and the aging process is that single gene mutations can be indeed the cause of accelerated aging and an enhanced occurrence of age-related disorders leading to a reduced individual life span (e.g. cancer). On the other hand, the downstream effects of such single mutations are again manifold and the view is manifesting that (1) normal (physiological) human aging is a multifactorial

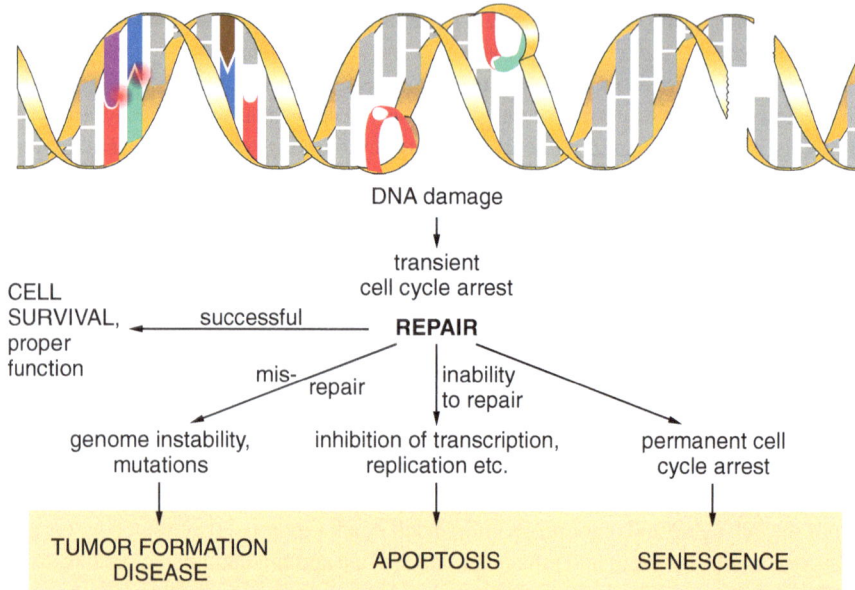

Fig. 3.7 Consequences of DNA damage on cellular level. The consequences of damage to the nuclear DNA largely depend on DNA repair efficiency: when repair is successful the cell can carry out proper function and re-enters regular cycling, in case DNA repair is not sufficient, tumor formation, apoptosis or senescence may be the consequence

process, simply because we are constantly affected by many different internal and external signals and (2) the maintenance of the genome is a key task of the cell and can decide between life and death.

As a reduced genome stability was identified as cause of an accelerated aging in humans (see progeria syndromes) there are also single genes and defined pathways that are responsible for an increased genome stability leading to a reduced aging and an increase in life span, at least in model organisms. A family of proteins called sirtuins is known to mediate such a genome stabilizing effect in yeast and nematodes (Howitz et al. 2003). After the presentation of DNA damage and DNA repair the focus now will be shifted towards defined genes that influence genome stability as well as life span.

3.3 The Sirtuins: The Ultimate Hope or Fallen Star?

Sirtuins comprise a whole group of genes and the family of sirtuin proteins in mammals consists of seven members. In the context of aging sirtuin 2 (Sir2) was initially described in yeast (Howitz et al. 2003) whereupon *Sir* stands for *silent mating type information regulator*. The mammalian homolog of Sir2 is named SIRT1. The protein Sir2 was identified as histone deacetylase. Histones are nuclear proteins that are overall positively charged, enabling the DNA which carries many negative charges to be tightly packed around them. Biochemically, this intimate connection is accomplished by electrostatical interactions between DNA (negative charges) and histones (positive charges). The positive charges of the histone proteins are caused by the presence of positively charged aminos acid residues (e.g. arginin, lysin; for details: Alberts et al. 2007). By enzymatic acetylation, the addition of acetic acid residues, the positive charges are neutralized and the histone-DNA packaging loosens up. The reversal of this chemical reaction, i.e. the removal of the acetyl groups, is called de-acetylation. In consequence, the histone-DNA packing again becomes very tight again. So, acetylation and deacetylation of the histones obviously affect the extent of the chromatin packaging and, therefore, also the function of DNA-based processes such as replication and transcription. In concert with the other sirtuin family members Sir2 activity is not restricted to DNA deacetylation but rather these enzymes target a wide range of cellular proteins in nucleus, cytoplasm and mitochondria that are post-translationally modified by acetylation (SIRT1, 2, 3, and 5) or ribosylation which is the chemical addition of ADP-ribose to a protein (SIRT4 and 6; Guarente 2011; Carafa et al. 2012).

Molecular mode of action of Sir2/SIRT1: Sir2 was discovered as prominent histone deacetylating enzyme and is one of a whole group of proteins that mediates transcriptional silencing by modulation of DNA packaging and is, therefore, a stabilizer of the genome. Strikingly, as experimentally observed, Sir2 activity can extend the life span of model organisms, such as yeast (*S. cerevisiae*) and *C. elegans*. In both organisms increased levels of Sir2, which can be rather easily introduced in these models by genetic manipulation and the generation of transgenic animals, lead to an increased life span. The activity of the sirtuins is dependent on the intracellular level of nicotinamide adenine dinucleotide (NAD), therefore, sirtuins are NAD-dependent protein deacetylases. NAD (correctly NAD^+) is a key co-enzyme for many different enzymatic reactions of the cellular metabolism. In addition to this role in redox-reactions, it is also a donor of ADP-ribose units in the process of ADP ribosylation (see Alberts et al. 2007). During the course of the energy metabolism NAD^+ is converted into NADH which is a so-called reduction equivalent that may ultimately deliver protons (H^+) to the respiratory chain reaction at the inner mitochondrial membrane. There, based on the accumulating proton gradient ATP synthesis is possible. In simple words: when the cells energy level is high, NADH is present in high amounts and the mitochondrial machinery generates a lot of ATP as the key energy currency of the cell. But under nutrient- and energy-wise more restricted conditions (caloric restriction, see below), NAD^+ is more prominent, allowing the reaction of sirtuins with NAD^+ and inducing sirtuin activity. Therefore, the activity of sirtuins is dependent on the actual energy status of the cell and sirtuins (hereby mainly speaking about SIRT1 in mammals) mechanistically link genomic stability, energy status and life span of the cell. After the first key findings on Sir2 (Howitz et al. 2003), this research field grew exponentially. The scientific attention and excitement was enormous because it appeared that with the sirtuins some first clear molecular switch was discovered that can directly regulate life span. It was believed that it is no longer

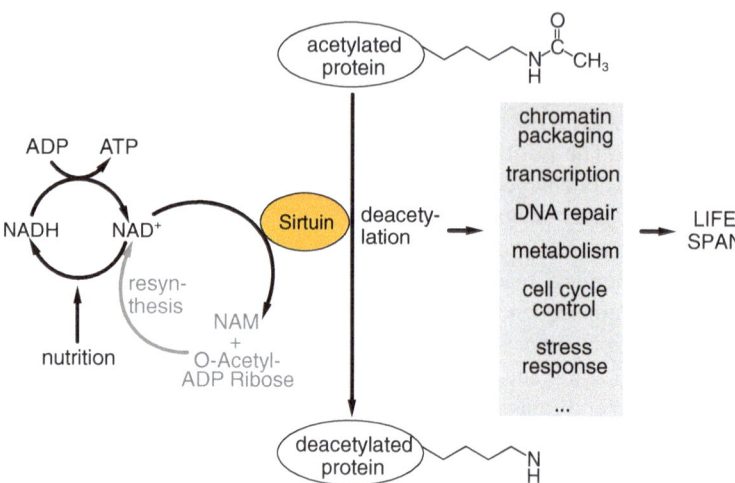

Fig. 3.8 Sirtuines and energy status. The deacetylation of various protein targets (e.g. histones) by Sir2/SIRT1 has several consequences including a closer histone-DNA interaction and a more tight chromation packaging. Moreover, the acetylation/deacetylation cycle affects transcription, DNA repair, metabolism, cell cycle control, stress response and, ultimately, the life span. Sirtuins are NAD+-dependent protein deacetylases and, therefore, depending on the energy level and the nutritional status of the cell

science fiction to find small molecules that by targeting the sirtuins may prolong life. Moreover, with that finding a molecular correlate was identified that might explain why caloric restriction may increase life span, namely via SIRT1 (Fig. 3.8).

Following their initial description sirtuin functions have been studied in many tissues and disease paradigms. In addition to a key role in aging, mainly SIRT1 was linked also to other conditions, such as metabolic and cardiovascular diseases and to neurodegenerative disorders (Guarente 2011). With respect to neurodegeneration, focus was put on Alzheimers disease (AD), the most important age-related degenerative and devastating deadly disorder. Here, one key pathological target, the biochemical processing of the amyloid precursor protein (APP), has been studied. And indeed, it was found that in transgenic mice overexpressing SIRT1 the production and deposition of a potentially toxic APP cleavage product, the amyloid beta protein (Aβ) was reduced (Donmez et al. 2010). Despite the still missing final experimental proof of the correctness of the *amyloid cascade hypothesis* in humans, the deposition of Aβ is believed to be a key pathological event causing AD and is desperately followed by academic researchers as well as by pharmaceutical industry for decades frequently ignoring other causal pathways involved in the pathogenesis of this multifactorial neurodegenerative disease (Dong et sl. 2012; Tam and Pasternak 2012; Masters and Selkoe 2012; Behl 2012). In addition, sirtuins have been linked also to the development of cancer based, for instance, on the observation that mice lacking SIRT3 (SIRT3 k.o. mice) are much more susceptible to mammary tumors compared to wildtype mice (Kim et al. 2010). Considering all this beneficial effects and modulatory activities on so many key processes linked not only to aging but also to disease, it was mandatory to search for molecules that can specifically upregulate sirtuin expression and then represent potential pharmaceutical drugs.

Stimulators of SIRT1: The main focus for sirtuin activators was put on compounds, ideally small molecules, for the stimulation of SIRT1. A highly interesting and extremely attractive chemical structure that was found to activate SIRT1 in a special screening protocol was resveratrol (Howitz et al. 2003). Resveratrol is a polyphenol bearing three phenolic groups in its structure. Its attraction comes from the fact that resveratrol is found in the extract of wine grapes, described in the 1970s. Promisingly, in the model of yeast cells resveratrol increased DNA stability and extended life span up to 70 % by stimulating Sir2 (Howitz et al. 2003). Another side aspect regarding additional activities of polyphenols such as resveratrol need to be mentioned here. Phenolic compounds including α-tocopherol (vitamin E) and 17β-estradiol (estrogen) at higher concentrations are powerful antioxidants, being able to protect cells under oxidative attack against oxidative damage and apoptosis (Moosmann and Behl 1999). These polyphenols, including resveratrol, even in low concentrations bind to estrogen receptors as shown in in vitro assays. Consequently, it needs to be considered that part of the described beneficial effects of resveratrol could be fully independent from its stimulating effects on SIRT1 and be due to other activities. Given that oxidative stress is a driving force of the degeneration of certain cell types in the body (e.g. neurons) and is acknowledged in the free radical hypothesis of aging (Harman 1956; see below) the antioxidant effects of certain compounds may be of some importance when addressing potential so-called anti-aging drugs. Subsequently, chemically distinct groups and even more effective activators of SIRT1 were identified. The assumption that resveratrol can extend the life span of model organisms by a defined molecular mechanism and mimics the effects of caloric restriction -meaning that a restricted nutrition (less calories) can lead to a longer life-made resveratrol and other small molecules highly attractive to be applied even in humans (Guarente 2011; Carafa et al. 2012; De Oliveira et al. 2012; Villalba et al. 2012; Villalba and Alcaín 2012).

So, for quite some time sirtuins were a prime target for the development and use of anti-aging drugs. Today the approach of targeting the sirtuins to influence life span is somewhat in doubt or at least in intensive and controversial discussion. Recent articles address the role of sirtuins in aging very critically (Couzin-Frankel 2011; Bourzac 2012). But what happened?

As pointed out in a *Science* news focus article in 2011, "Work that pinpointed the control of aging to a handful of genes is being taken apart by some of the scientists who made early discoveries" (Couzin-Frankel 2011). Obviously, some of the experimental work trying to extend life span by targeting the sirtuins in fruit flies and *C. elegans* could not be reproduced and the possibility of actually transferring the findings from model organisms to mammals and humans is strongly in doubt. The fact that the scientists that took part in the initial discovery and development of the link between sirtuins and aging are now arguing against each other strongly suggests that caution should be given when estimating and rank sirtuin function in aging. The sirtuin field is presently highly polarized and "sirtuins are still controversial"

(Bourzac 2012). A personal note: also the fact that the topic of anti-aging is a billion dollar market and from early on pharmaceutical companies partly owned by the researchers are involved may explain why the sirtuins started as such a great hope and may eventually end as fallen star. Very recently the link between sirtuins and caloric restriction was strengthened by addressing new data showing a systematic influence of diet-associated sirtuin activity on mammalian physiology (Guarente 2013). But still, as intriguing the view is that aging is controlled by a single set of genes or even one single gene that can be manipulated, after following the sirtuin case one has to realize and stress again that the process of aging is much more complex and that there probably is no such thing as a master switch of aging. Nevertheless, looking at some experimental facts dealing with measures to extend life span and when reviewing the literature on experimental possibilities to manipulate life span of organisms, it is mostly accepted that a controlled dietary restriction and the reduced activity of nutrient-sensing pathways (e.g. the insulin pathway) slow aging by similar mechanisms, which are evolutionary conserved (Fontana et al. 2010). Therefore, the process of caloric restriction and its link to aging should continue this discussion.

3.4 The Impact of Caloric Restriction on Life Span

Luigi Cornaro, an Italian philosopher and individualist who lived in Venice from 1467 to 1565 summarized in some autobiographic work called *Discorsi della vita sobria* (Vom mäßigen Leben, About a moderate life) written at the age of 83 some reflections why he actually got that old. He presumed that his old age and rather good health is the consequence of a strict diet. It is traditional knowledge that Cornaro ate only the minimum he needed and that he always thoughtfully selected his food (Cornaro, in translated form published in 1903). He finally died at the age of almost 100 years which was for the epoche of Renaissance a rather old age. In the literature of the last century the general concept of altered nutrition, mainly meaning a reduced dietary and nutritional intake, and its link to organism aging was suspected and reported as early as in the 1930s by the gerontologist and nutrition scientist Clive McCay (McCay 2000). He in fact for the first time suggested that a 30 % reduction in dietary intake can be a possibility to extend life span without negative effects on general nutrition (malnutrition) employing rodents as models. As many subsequent studies indeed confirmed an extension of life span upon caloric restriction in other species including nematode worms (*C. elegans*) and fruit flies (*D. melanogaster*), it became apparent that a reduced caloric intake over time, on top prevented also age-related dysfunction and disease such as cancer and diabetes in mice (Fontana et al. 2010). Based on that it can be argued that, very likely, the extension of life was a result of a decreased disease incidence and of less disease-related changes that affect cellular and organ activities leading to loss of function and death.

"Caloric restriction is the most powerful known intervention in aging" (Luigi Fontana): Recently, studies on the effect of caloric restriction on life span in various non-human species were discussed by several opinion leaders in aging research.

In addition to the direct supply of cells and organisms with nutrients, obviously also the molecular pathways that sense extracellular nutrients (mainly glucose) and that serve as signal transductors can modulate the life span. It has been shown that alterations in nutrition-sensing pathways caused by mutations or provoked by chemical inhibitors can mimic caloric restriction. Two key nutrient sensors are currently focused on (1) the family of insulin and insulin-like growth factors I and II (IGF-I and IGF-II) and their membrane receptors and (2) TOR protein (TOR stands for *target of rapamycin*). In their seminal review on this very topic the authors Luigi Fontana, Linda Partridge and Valter D. Longo summarize that "dietary restriction and reduced activity of nutrient-sensing pathways may thus slow aging by similar mechanisms, which have been conserved during evolution" (Fontana et al. 2010). Based on all these results studies were launched applying this concept in non-human primates to finally bridge the gap between model organisms and humans. In fact, first results of two still ongoing long-term studies on caloric restriction have been reported in 2009. A population of rhesus macaques at the Wisconsin National Primate Research Center lived under a moderate (meaning approximately 30 %) caloric restriction and showed a significantly lowered incidence of aging-associated deaths. At the reported time point 50 % of control fed animals with ad libitum food access survived but 80 % of those monkeys under caloric restriction. As in other mammals observed (e.g. mice), reduced food intake resulted in a delayed onset of different age-associated pathological changes, including a reduced incidence of diabetes, cancer, cardiovascular disease, and brain atrophy. Although the maximum life span or a possible extension of it was not observable yet due to the design of the study the extension of life in a majority of the monkeys under reduced food allow the interpretation that this intervention can slow aging in a primate species (Colman et al. 2009) (Fig. 3.9).

An independent second study was also conducted in rhesus monkeys, this time at the US National Institute on Aging (NIA). Interestingly, the final outcome with respect to life span extension by caloric restriction was quite different. The NIA study summarized that caloric restriction (meaning a reduction of intake of nutritious diet by 10–40 %) in young and older age rhesus monkeys did not extend life span but improved the general survival outcomes. Nevertheless, this study also found beneficial effects of reduced food intake on cancer rates (Mattison et al. 2012). So while the Wisconsin study can be interpreted to demonstrate a strong link between health benefits and the extension of life span, the NIA study rather suggests a separation between health effects, morbidity and mortality. These obvious inconsistencies may be explained mainly by the differences in the respective study design. Importantly, while the monkeys of the control group in the Wisconsin study were completely unrestricted with regard to nutritional intake and actually could have as much food as they wanted (*ad libitum*), the control group monkeys in the NIA study received a specially prepared and healthy diet. Also, a full evaluation of the data can only be performed when the studies have finally been finished which will, due to the life expectancy of this non-human primates, take at least another 10 years. Nevertheless, the punch line result of both studies already is that less food was beneficial for the general health of these primates.

Fig. 3.9 Effects of caloric restriction on the appearance of rhesus monkeys. Rhesus monkeys on a nutritious, but reduced calorie diet (*left*) and on an unrestricted diet (*right*) pictured at the Wisconsin National Primate Research Center at the University of Wisconsin-Madison on May 28, 2009 (photo courtesy of Jeff Miller / University of Wisconsin-Madison)

Based on the experimental animal studies and the presentation of their preliminary results in the media different groups of people have decided to follow such a caloric restriction approach and to adjust their eating behavior, with *Caloric Restriction with Optimal Nutrition* being the keyword. Persons that live under such a diet are under medical observation and researchers as Luigi Fontana report on great effects of a for instance, 8 year long ongoing caloric restriction of the human body on the cardiovascular system (Weiss and Fontana 2011). As observed in animal models caloric restriction changes the general hormone status and has a strong impact on the metabolism also in humans (Heilbronn et al. 2006). It is acknowledged that a long term caloric restriction is preventive against Diabetes mellitus type II, arteriosclerosis and hypertension. Seen from the other side, under the extreme nutritional conditions as in obesity due to the general health status the medium life expectancy is shortened. Regardless of the multiple effects of a reduced (and balanced) diet on the whole human body, its various tissues and cell types, it is way too early to pronounce caloric restriction as a possible fountain of youth. Needless to say that at this state of experimental findings and knowledge it is of great importance to study the effects of a reduced caloric intake at the molecular level to understand the exact underlying mechanism. It has to be stated that despite the fact that caloric restriction improves the health of lab animals and humans, reduces age-associated detrimental pathologies and may (as a direct consequence or independently) extend life span, at this point it is still open what the exact molecular mechanism of caloric restriction is and how it actually works. Here, a look into studies on *C. elegans* and other model organisms gives some insight.

3.5 Genes that Extend Life in Animal Models

Age-1, Daf-2, Daf-16 Genes in *C. elegans* are Linked to Nutrition: The work of Tom Johnson, David Friedman, Cynthia Kenyon, Adam Antebi, Gary Ruvkun, Pam Larson, Linda Partridge and many others (the authors are fully aware that the list is not complete!) contributed and showed that single genetic changes can modulate the life span of model organisms. These findings were achievable since well characterized and less complex model organisms such as the nematode *C. elegans* were employed. In fact, the progress in aging research and the understanding of age mechanisms were and still are strongly driven by investigating *C. elegans*. This approximately 1 mm long and transparent worm consists of around 1,000 cells. His life cycle is short, reaching the adult stage after a development phase of 3–4 days. Its life span is usually 3–4 weeks. Actually, focussed research on the molecular biology of *C. elegans was* started by Sydney Brenner in 1974. Since then this worm is one of the key model organisms for molecular medicine research, including aging research. The cellular networks of the different worm tissues are well studied and so are key genes. To date, over 100 genes are known that when manipulated can change the life span of this animal. Large research communities and consortia focusing on *C. elegans* are working closely together. Collections of gene and gene expression libraries and a large number of *C. elegans* strains with genetic variations are available and accessible to the research community strongly speeding up the research progress in the field.

As early as in 1988 it was described that a special *C. elegans* strain called *age-1* showed an increased life span when compared to the other worms observed on a Petri dish. Age-1 mutant worms lived approximately 60–80 % longer as the wildtype and, therefore, age-1 was the first so called longevity gene (Friedman and Johnson 1988). This early hallmark discovery strongly promoted the view that the life span and the aging process of a full organism can be modified (delayed and perhaps accelerated) by single genes. This observation marked quite a shift of paradigms since it argued against the view that the aging process is of multi-genetic and multi-factorial origin that already at that time was highly accepted and currently again is. With the strong technological developments in modern molecular biology coming up the subsequent molecular and functional analysis of age-1 was possible. Cloning and characterization of the age-1 gene revealed that it codes for an enzyme called phosphatidylinositol-3-OH kinase (Morris et al. 1996). This enzyme is part of an intracellular signal transduction cascade activated after binding of insulin to its receptor at the cell membrane, downstream ultimately leading to transcriptional changes. Another breakthrough in *C. elegans* aging research was reported by Cynthia Kenyon. She identified a second gene that when mutated leads to a doubling

Fig. 3.10 Life span modulation in *C. elegans*. The worm line daf-2 (mutation in the insulin/IGF-1 receptor gene) has an extended life span whereas the daf-16 line (mutation in the gene coding for the transcription factor DAF-16 downstream of the insulin signaling) displays a reduced one (graph courtesy of Andreas Kern, Institute for Pathobiochemistry, University Medical Center Mainz)

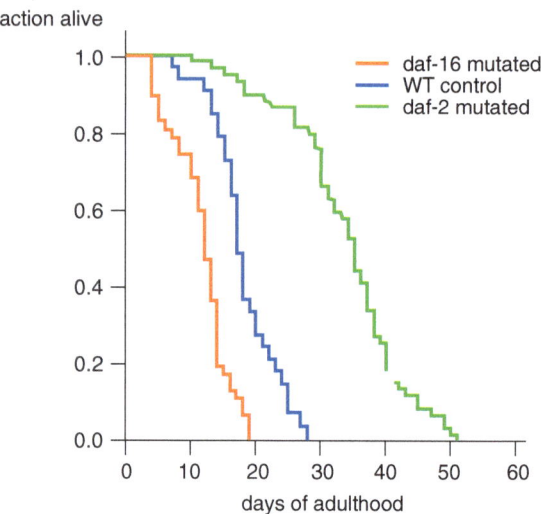

of the life span of the worm, called *daf-2*. The correspondent genetically altered worm line exhibited the largest reported life span extension in any organism at that time (Kenyon et al. 1993). Moreover, in this 1993 paper it was made clear that this life span extension requires the activity of a second gene, called *daf-16* (Fig. 3.10). Daf-2 and daf-16 are functionally related genes. In a subsequent study daf-2 and age-1, both having this dramatic effects on the life span of *C. elegans* were directly compared.

Age-1 and daf-2 mutations cause the same effect on life span and do so by affecting similar processes (Dorman et al. 1995). These first data suggested that there could be something like a common downstream pathway that regulates aging. The next step, of course, was to decipher which actual proteins are encoded. Interestingly, also daf-2 coded for a protein involved in the signaling processes that are activated after binding of insulin, namely the receptor for IGF-1. So, whenever insulin or IGF-1 activate cells the proteins coded by age-1 and daf-2 mediate the intracellular insulin/IGF-1 signaling (Kimura et al. 1997). Interestingly, daf-16 acts in concert with age-1 and daf-2 and daf-16 was characterized as a factor directly regulating transcription. This transcription factor was called FOXO that binds to the DNA when the phosphoinositol-kinase coded by age-1 upstream in the signaling cascade is activated. Today we know that FOXO as the downstream executor of upstream signaling events induced by insulin and IGF-1 plays a role far beyond the modulation of the aging process (Salih and Brunet 2008). The so called forkhead box, class O (FOXO) belongs to a large family of forkhead transcription factors that are characterized by a conserved (forkhead) box DNA-binding domain and DAF-16 is the only ortholog of the FOXO family in *C. elegans* (Fig. 3.11).

Taken together, all this work was suggesting that signals from outside the cell, actually a hormone (insulin) as soluble factor, can regulate the aging process. A muta-

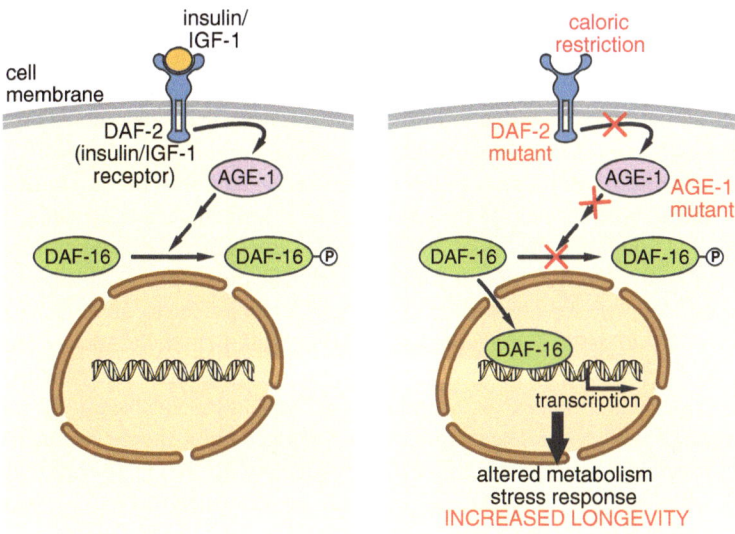

Fig. 3.11 AGE-1, DAF-2, DAF-16 as part of the insulin signaling pathway. Age-1, daf-2, and daf-16 represent *C. elegans* strains with a modified life span. The proteins encoded by these genes interfere with the insulin/IGF-1 signaling pathway; upon stimulation the insulin/IGF-1 receptor activates the phosphatidylinositol-3OH kinase AGE-1. Via several steps the transcription Factor DAF-16 gets phosphorylated leading to its retention in the cytoplasm. This pathway can be disrupted at different sites (shown in *red*) (modified after Ewbank (2006); whereas in worms and flies there is only one type of insulin/IGF-receptor known, in mammals two different receptors are described, the insulin receptor and the IGF-1 receptor)

tion in the daf-2 gene, which represents the insulin/IGF-1 receptor in the nematode model, leads to a dysfunctional receptor. Consequently, there is simply no signal getting into the cell and no activation of the subsequent cascade in which daf-16/FOXO is involved. Of course, it was known by then that insulin levels are increasing in the body when the blood glucose levels are high as a consequence of nutritional intake. It is one key function of insulin then to mediate and actually enhance the glucose uptake in special tissue. In humans insulin-dependent glucose transporters are expressed in fat cells (adipocytes), skeletal and heart muscle cells. Uptake of nutrients (mainly glucose) drives the metabolism, an interruption of this metabolic processes, e.g. caused by a defect insulin receptor as in the case of daf-2 mutants, leads to decreased glucose uptake just as well. Such a decreased glucose uptake during starvation (no food intake) or in the course of caloric restriction (restricted food intake) also leads to a decreased insulin-dependent intracellular signaling. In simple words one can conclude that here the concepts of caloric restriction and of defective insulin signaling in mutant worm lines converge and have basically the same end result: an increased life span in model organisms. This interpretation is very interesting and also intriguing since it implies that extracellular hormones and potentially also pharmaceuticals that mimic or counteract hormonal activities can influence the aging process. But in the next step it was important to investigate whether the modulation of the aging process as uncovered in *C. elegans* can also be reached in other species and, most importantly also in mammals. Initially, it was the work of Linda Partridge and of Marc Tatar that followed such questions in another model organism, the fruit fly *Drosophila melanogaster* (*D. melanogaster*).

Life span modulation and transfer of knowledge from worms into fruit fly and mouse: Like *C. elegans* the fruit fly animal model is well characterized, cells, tissues, networks and regulating processes and molecules are studied based on pioneer work employing many different mutant fruit fly lines. The development of this organism also is well known; for Drosophila the larval development to an adult organism takes about 10–12 days. The regular and full life span of *D. melanogaster* under lab conditions is between 50 and 80 days and is, therefore, ideal to study medium and maximum life span and life span manipulations. Due to a lack of space, here, the story of the findings in Drosophila can not be acknowledged but it should be stated that in 2001 Partridge and Tatar reported in a side-by-side publication in *Science* that based on work with mutant fly lines the life span in *D. melanogaster* (as in *C. elegans*) can indeed be modulated by interfering with the flies insulin signaling pathway (Clancy et al. 2001; Tatar et al. 2001). With these findings it was clearly shown that in two distinct species a decrease in insulin/IGF-1 signaling leads to an extension of the life span indicating also the evolutionary conservation and importance of this molecular link between the nutritional status (e.g. caloric restriction), nutritional signaling and aging. The next logical step was the proof of applicability of the observed molecular link in the mammalian organism.

Worms and fruit flies have only short life spans, mice on the other hand live approximately 2–3 years under lab conditions. When using mouse models in medical research, of course, the main goal is getting closer to the human organism. With the advent of recombinant mouse technologies and the development of experimental mouse genetics a huge amount of mouse models targeting genes linked to human disease were generated. As interesting and winning the work with mouse models in general is, it is still a great controversy going on, whether a mouse can represent a good model system for human disease. Just recently, a study has focused on the relevance of mouse models of inflammation and has again raised great doubts concerning the transferability from mouse to man (Shay et al. 2013). Needless to say that this should be taken even more into account when employing mice in aging research since a maximal life span of 2–3 years is not comparable to a maximal life span of 100–120 years of humans. But mice are mammals and many aspects of the murine metabolism and physiology are smiliar to some extent in humans and even more as there is a huge basic genetic homology between mouse and man.

In the 80s, it was the work of Andrzey Bartke that identified the so-called "Ames dwarf" mice showing an increased life span of approximately 50 %. These mice carried a mutation in a gene called Prop1, representing a transcription factor involved in the development of the hypophysis. This brain areal is an important part of the mammalian hormonal signaling axis and is connected upstream to the hypothalamus (another key regulatory brain region; endocrine hypothalamus-hypophysis axis) and initiates the expression of many target genes in various target tissue (e.g. testis, ovaries, thyroid gland). One main task of the hypophysis is the secretion of peptide hormones that enter the blood stream and bind to membrane receptors in target tissue throughout the body. The Ames dwarf mice, therefore, displayed a variety of hormonal changes and reduced levels of, for instance, the hypophysis-derived growth hormone. In addition and interestingly, these mice had decreased serum levels of IGF-1 and it was known that one activity of the growth hormone was a regulatory influence on IGF-1. So, obviously, also in mice some link of longevity to the insulin-signaling processes exists (Brown-Borg et al. 1996). Of course, the story further developed and is elegantly addressed in various more recent expert reviews (Bartke 2011; Yakar and Adamo 2012; Brown-Borg and Bartke 2012). Additional work from other laboratories confirmed and extended the link of longevity in mice to insulin signaling focusing on the insulin and IGF-1 receptors, for instance by genetically knocking out the insulin receptor exclusively in fat tissue leading to lean mice with a long life span (Holzenber et al. 2003; Blüher et al. 2003). A further expansion and support of the molecular and mechanistic link between energy metabolism and aging was given by the discovery of the gene klotho, coding for a specific transmembrane protein that is functionally indirectly linked to the insulin pathway. Transgenic mice overexpressing klotho show an increased life span, mice lacking klotho develop premature aging. Furthermore, it was found that the family of Klotho proteins may act as important co-receptors for FGFs, the fibroblast growth factors that are involved in the regulation of a broad range of processes in the metabolism as excellently reviewed recently (Razzaque 2012; Kuro-o 2012).

To sum it up, besides many experimental studies that have been presented since the initial description of the link between caloric intake, metabolism and aging (for review: Fontana et al. 2010)—and there are still many aspects that cannot be addressed here (e.g. Yin et al. 2013)—and despite first findings in rhesus monkeys the final transfer of the food intake-life span-link into the human situation was not yet possible. As outlined above and to stress it again, there is no doubt that a reasonable restriction in food intake is good for the general health status of humans. Of course, it is a personal decision of any individual to follow such a nutritional strategy to what extent ever. And with almost all things in modern life, one may find help and suggestions in the internet (e.g. http://www.crsociety.org) and may even join professional, semi-professional, stronger orthodox or less orthodox communities. Having obvious beneficial effects of caloric restriction in mind, nevertheless, a long-term restricted diet may have general side effects as observed in highly underweight individuals. Moreover, one should also think back to the strategy of the ancient Coronaro who not only ate less food but also carefully selected his nutrition.

But also the influence of even rather short periods of caloric restriction on the human body should be seriously considered. One example and a final thought on the benefit of caloric restriction should focus on the brain and cognitive performances. Based on animal studies that clearly demonstrated that low calorie diet and enrichment of food with unsaturated fatty acids can be beneficial for cognitive activities in aged animals, a prospective interventional study in humans was performed (Witte et al. 2009). Fifty healthy subjects, normal weight to overweight elderly people were included and divided into 3 study groups: (1) the caloric restriction group (30 %

reduction of food intake), (2) the group with normal food intake plus food supplementation with about 20 % more unsaturated fatty acids, and (3) the group of
control individuals with no change in their nutritional intake. After 3 months cognitive functions, in particular, memory performance was assessed and it was found
that the group experiencing caloric restriction showed a significant increase in verbal
memory scores. This memory improvement could be correlated with decreases in
baseline fasting plasma levels of insulin and high sensitive C-reactive protein, most
pronounced in subjects with best adherence to the diet. The two other groups showed
no significant memory changes (Witte et al. 2009). While the exact mechanisms
underlying this significant improvement of memory performance are not yet clear
and hard to reveal, these observations impressively show that the general metabolism
status of the whole body can directly modulate (and in this experimental paradigm
improve) the fine-tuned neuronal homeostasis in the human brain. Consistently, in
an interesting recent study employing a mouse model of Alzheimers disease it was
shown that also hunger in the absence of caloric restriction diminished the levels of
Aβ and inflammation indicative of microglial activation at 6 months of age compared
to the control group, similar to the effect of CR and improved cognition (Dhurandhar
et al. 2013). Thus, similar to the short-term overall caloric restriction in humans, also
a repeated application of a protein intake diet impacts on brain function. Moreover,
the recent study again connects a restriction in dietary uptake mechanistically to the
IGF-1 system and to significant health benefits that may eventually lead to life span
extension. Widening this view and in simple words: the brain is attached to the body
and if you are doing good to your body you are also doing good to your brain.

Nutrition sensor mTOR (mammalian target of rapamycin): Different lines of
evidence indicate the central role for insulin/IGF in mediating the effects of caloric
restriction on longevity. Another key control protein that has a link to insulin/IGF signaling in the cell, is the protein TOR (target of rapamycin). Rapamycin is a compound
that was first isolated from the bacteria *Streptomyces hygroscopicus* found for the
first time on the Island of Rapa Nui (Easter Island) giving the bacterial compound
its name. The protein mTOR (mammalianTOR) is a large serine/threonine kinase
and belongs to the phosphatidylinositol 3 (PI3-) kinase-related protein kinase family. More exactly, mTOR represents the catalytic subunit of two protein complexes
mTORC1 and mTORC2. Rapamycin targets mTORC1 und inhibits its downstream
signaling activity. Interestingly, an inhibition of mTOR blocks the activation of a
cdc-kinase and the subsequent complexation with cyclin E keeping cells in the G1-
phase and preventing transition to the S-phase of the cell cycle (Foster et al. 2010;
Dobashi et al. 2011) Fig. 3.12. For its influence on the cell cycle of proliferating
cells rapamycin is used as an anti-cancer drug and it is also known as immunosup-
pressant in man. The inhibition of mTOR by rapamycin then leads to autophagy
induction. Hence, mTOR is an upstream regulator of an evolutionary highly conserved process called autophagy (auto/self, phagy/eating), a cell-intrinsic degradation
process occurring in membrane compartments with high lysis activity, the so-called
lysosomes. By the process of autophagy cells can self-digest intracellular waste such
as damaged and dysfunctional organelles (e.g. mitochondria; this autophagy process
is then called mitophagy), aberrantly folded and/or modified dysfunctional proteins or

protein aggregates (Chen and Klionsky 2011). Therefore, autophagy is a central part of the cellular protein and organelle quality control system that ensures the intracellular homeostasis. The first to describe an intracellular process by which cells can digest their own cytoplasmic material within the lysosome compartment were Christian De Duve and his group (De Duve and Wattiaux 1966). Another part of the protein quality control is the proteasome, a multienzyme complex that degrades proteins that are enzymatically labeled for that purpose (enzyme: ubiquitin ligase) with a specific ubiquitin protein tag, since all proteins inside the cell have a defined half life. An important requirement for protein degradation via the proteasome is that proteins can be unfolded so that they can physically enter the proteasome degradation machinery (Mogk et al. 2007; Fredrickson and Gardner 2012; Amm et al. 2013). Aggregated proteins cannot unfold and need to be removed by an alternative pathway, the autophagy. As mentioned autophagy is evolutionary highly conserved and one main inducer of this process in all the species and cell systems investigated so far is starvation, deprivation of nutrients in the extracellular compartment and extracellular stress. When due to starvation no amino acids are supplied from the extracellular compartment an enhanced turnover of proteins is supplying the cell with amino acids as protein breakdown products.

Autophagy is tightly regulated and follows a canonical process involving many proteins and regulators and a key upstream control protein which is mTOR. Mammalian TOR integrates various input signals (e.g. insulin, growth factors), senses nutrients, energy and oxygen levels. Since autophagy is of such key importance for the cellular function and survival a disturbance of this process may lead to a disruption, for instance, of the protein homeostasis which is closely linked to aging and age-related disease including AD (Morawe et al. 2012). And, regarding aging, in model organisms such as yeast, nematodes, flies and even in mice life span can be prolonged by the induction of autophagy. Autophagy induced by pharmacological compounds, most prominently rapamycin, extends life span but also protects against protein aggregation-associated dysfunction in transgenic animals expressing aggregation-prone disease-associated proteins such as the Huntingtons disease causing protein huntingtin (Hochfeld et al. 2013). Mice at the age of 600 days (which is frequently compared to a human age of approximately 60 years) that were treated with rapamycin displayed an increase of the life span, about 14 % in females and 9 % in males (Harrison et al. 2009). So, even a rather late start with such an intervention still leads to a significantly extended individual life span. Due to many side effects, including cateracts and immune suppression-associated syndroms, rapamycin is definitively not a candidate for testing in humans with regards to delay human aging. On the other hand it shows the potential the targeting of the autophagy pathway may have, even in the mammalian organism. But up to date it is not clear how exactly rapamycin mediates this effect in mice, whether by counteracting and preventing age-associated disease or by a direct interplay with the aging process or by a combination of both. In their original paper the authors state "To our knowledge, these are the first results to demonstrate a role for mTOR signaling in the regulation of mammalian life span, as well as pharmacological extension of life span in both genders. These findings

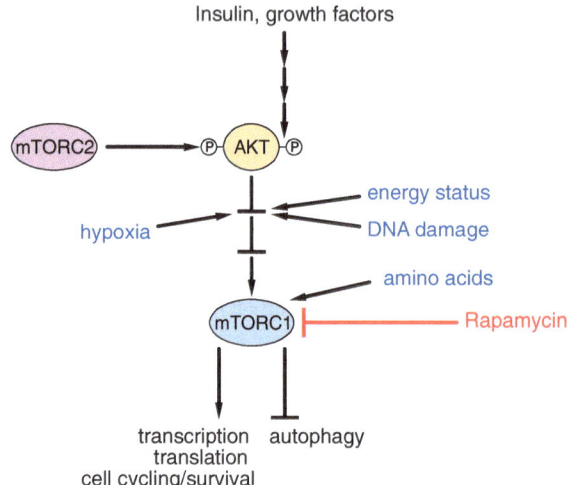

Fig. 3.12 Mammalian target of rapamycin, mTOR signaling. Mammalian target of rapamycin (mTOR) represents the catalytic subunit of two protein complexes mTORC1 and mTORC2, rapamycin targeting only mTORC1 und inhibiting its downstream signaling activity. mTORC1 integrates a variety of intra- and extracellular signals and targets key cellular processes. In its active form mTORC1 stimulates transcription and translation thereby regulating cell growth and blocks autophagy. mTORC2 affects mTORC1 via the activation of the protein kinase Akt

have implications for further development of interventions targeting mTOR for the treatment and prevention of age-related diseases" (Harrison et al. 2009) (Fig. 3.12).

Although the experimental findings on the impact of the modulation of mTOR and the autophagy pathway cannot be translated into humans, recently interesting observations were reported on mTOR and life span in humans. By studying the expression of mTOR and associated genes in tissue samples of people in a special aging cohort (Leiden Longevity Study) the mRNA levels of mTOR signaling genes of nonagenarians, meaning people in their tenth decade, and middle-aged controls were investigated. Shortly, it was found that old age is associated with a differential expression profile of mTOR pathway genes. One of these is the RPTOR (Raptor) gene (Passtoors et al. 2013), the raptor protein being a central component of mTORC1. All this together strongly suggests that mTOR-regulated processes in addition to modulating disease and aging in a variety of animal models also may have an influence on human aging. These results should encourage more research on this potential interplay but one should always have in mind that autophagy regulated by mTOR is a multifunctional process, targets various downstream processes,and is not easy to address pharmacologically without untoward effects (see rapamycin).

3.6 Free Radical Theory of Aging

So far the focus on possible mechanisms and regulators of the aging process has been put on defined genes and intracellular pathways that could be directly linked to a modulation of the life span of organisms. Manipulation of certain genes (e.g. daf-2) as well as intervention with a distinct intracellular pathway (e.g. mTOR signaling) can affect the life span of model organisms. On the other, we learned that simply a reduced caloric intake may extend life span as well. One will find a link, too, between life span extension under caloric restriction and the next aging theory that will be presented here, the free radical theory of aging. The reason is that a lower mitochondrial energy metabolism (as under caloric restriction or starvation) consequently will also lead to a decrease in mitochondria-derived free radical generation and subsequent oxidative stress (Pamplona and Barja 2006).

Initially formulated in 1956 by Denam Harman (Harman 1956), the *free radical theory of aging* actually survived already decades. This view on aging is based on the fact that biomolecules are challenged over (life) time continuously by many exogenous and endogenous factors and, therefore, experience a significant amount of wear and tear driven by oxidants. The structure of the cellular components can be chemically altered by oxidations and, ultimately, an altered structure means very often reduced or loss of function or, in some cases, even gain of toxic function. All three types of key biomolecules, proteins, lipids, and DNA can be chemically modified and, in general, oxygen attacks all molecules in the cell. Oxygen is of central importance for our life but it may also cause (oxidative) harm. In mammals over 90 % of energy is generated by the chemical reduction of molecular oxygen (O_2). During the respiratory chain in the mitochondria oxygen is reduced to water (H_2O; Fig. 3.13). Thereby, not only energy is produced but toxic side products of O_2 are formed as well. Such toxic side products are free oxygen radicals, also called reactive oxygen species (ROS) that are the molecular mediators causing *oxidative stress*. In fact, proteins, lipids and DNA are highly susceptible to oxidations and to such oxidative stress that permanently attacks these molecules in the cell. But, what does oxidation of a molecule mean? What is oxidative stress exactly? How can it do harm to the cell, change its function or lead to cell death? And, finally, what has this to do with the aging process?

By definition, *oxidation* means a chemical reaction in which an atom, an ion or a molecule gives electrons to its reaction partner. This electron transfer changes the chemistry of the electron acceptor molecule. A stable atom or molecule that accepts an additional electron (and then carries an unpaired electron) is called a *radical* and can become highly reactive because it has an intrinsic need to loose and transfer this additional electron as soon as possible which is the basis for its reactivity (*reactive radical*). Oxygen is present in high amounts in the mitochondria to be reduced to water by a stepwise *reduction (* the opposite reaction of oxidation) by reacting with electrons and hydrogen. During this process very often O_2 accidently receives additional electrons becoming itself a radical ($O_2^{\cdot-}$, superoxide radical) and is then the origin of additional oxygen derived radical species.

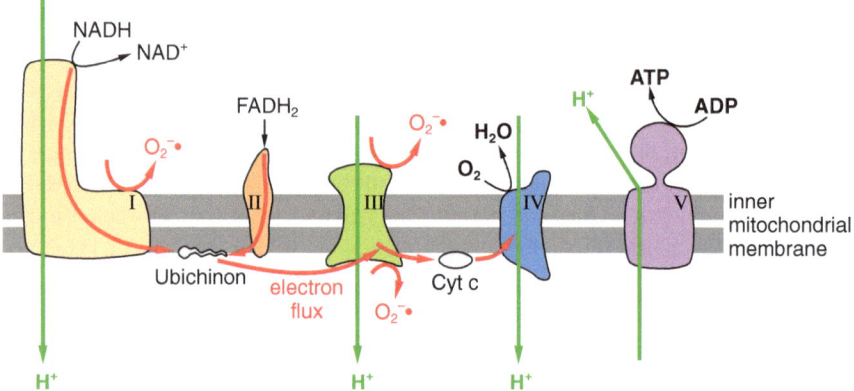

Fig. 3.13 Mitochondrial respiratory chain: the reduction of oxygen to water. In the inner mitochondrial membrane through a complex sequence of reactions molecular oxygen (O_2) is reduced to water (H_2O) as electrons are transferred via interaction with four integral membrane protein complexes (complexes I–IV). Complex V is also known as ATPase and is the enzyme system that leads to energy generation in form of ATP. The driving force of the ATPase is a proton (H^+) gradient that builds via the accumulation of reduction equivalents (NADH and $FADH_2$). During the electron passage O_2 frequently receives an extra electron and O_2^-, the superoxide radical is formed (for simplification unidirectional arrows are depicted; modified after Müller-Esterl 2011)

While the reduction of O_2 to H_2O in the respiratory chain is the desired physiological reaction, the production of superoxide radicals O_2^- can be considered as a reaction accident. In addition to the respiratory chain in the mitochondria O_2^- is also generated by various cell-intrinsic enzymatic systems, in these cases on purpose. Oxygen radicals have also regulatory functions, for instance, in the redox-regulation of different proteins and in the modulation of the transcription of oxygen-responsive genes. Moreover, in specialized immune cells (phagocytes) of the mammalian immune defense system against infections the immense reactive and also destructive power of O_2^- is used to oxidatively destroy exogenous particles and microbes and to kill phagocytosed microorganisms (Nauseef 1999).

A major breakthrough in free radical research was the discovery of the enzyme superoxide dismutase (SOD) that converts superoxide radicals to oxygen and hydrogen peroxide (H_2O_2) by McCord and Fridovich (McCord and Fridovich 1969, 2013). Based on these and other findings great attention was put on the mitochondria as ROS generating systems and free oxygen radicals as metabolic byproduct of the respiratory chain. In 1972 Denham Harman delivered a reevaluated and updated version of his initial free radical theory of aging considering the novel findings in free radical biochemistry (Harman 1972); a personal review on all significant developments has been presented by him also (Harman 2009). The free radical theory of aging is highly comprehensible since while living under oxygen conditions our cells and tissues are

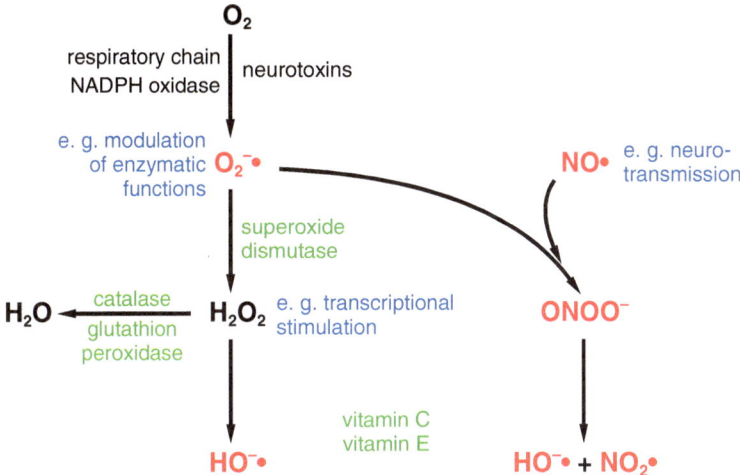

Fig. 3.14 Reactive oxygen and nitrogen radical species, ROS and RNS. ROS and RNS (*red* writing) are formed permanently and are interconnected. Besides their high reactivity potential they serve also vital cellular functions (*blue* writing). Antioxidant enzymes and antioxidants (*green* writing) can detoxify the ROS. A lack of detoxification and/or an overboarding production can lead to ROS and RNS accumulation. (Modified after Behl and Moosmann 2008)

permanently challenged by oxygen and oxygen-derived radicals. Subsequently many lines of experimental evidence supported (and still support) this theory. Basically, some key evidence of the plausibility of this aging theory should be mentioned here: (1) oxidative damage to cells and tissues is found to be increased with age of experimental animal models, (2) different conditions that extend life span also reduce the overall oxygen free radical formation and subsequently occurring damage, (3) certain genetic manipulations that extend life span of animal models are directly associated with lower oxidative damage (Halliwell and Gutteridge 1999).

Reactive oxygen species, reactive nitrogen species and oxidative stress: By mistakes in the transfer of electrons during the respiratory chain in the mitochondria molecular oxygen can be charged with an extra electron leading to the superoxide radical (O_2^-). This superoxide radical then is the origin of other radicals and oxidizing molecules that are addressed as reactive oxygen species, ROS. As shown, ROS are generated primarily at the respiratory complexes I and III and are highly reactive. The most prominent representatives are the already mentioned superoxide radical (O_2^-) and hydrogen peroxide (H_2O_2). Hydrogen peroxide by its chemistry is not a radical itself but is the origin for the generation of the highly reactive hydroxyl radical (HO·). The ROS differ in the degree of their reactivity and, therefore, also in their reactivity range inside (or outside) cells. The hydroxyl radical is of highest reactivity and is oxidizing the biomolecules in the immediate environment. Moreover, peroxynitrite ($ONOO^-$) that is generated when nitric oxide (NO·) is reacting with superoxide radicals (O_2^-) is highly reactive. Peroxynitrite and the less reactive nitric oxide belong to the family of reactive nitrogen species, RNS (Fig. 3.14).

Taken together, a variety of ROS and RNS can be formed inside the cell. Although the mitochondria are addressed as primary source of ROS, also outside of these organelles ROS are generated. One enzymatic system that can directly produce superoxide radicals in the cytoplasm is the NADPH oxidase (Bae et al. 2011).

ROS may have also signaling functions and no destructive activities at all. They may act as oxygen and redox sensitive modulators of transcription as very clearly shown for hydrogen peroxide, an activator of the transcription factor NF-κB (Li and Karin 1999). So radicals show a janus face. They are representing key intracellular signaling factors but they also can destroy the cell. Here, of course, the concentration in which the ROS are generated and are eventually accumulating is of critical importance. This basically also applies to RNS but nitric oxide (NO·) is of low reactivity so that it is used also as signal messenger from cell to cell, even in the central nervous system. Indeed, for nerve cells NO· is an important messenger in synaptic neurotransmission and has, therefore, been shown to play a role in the formation of memory in the mammalian brain. Originally, NO· has been identified as *the* locally acting factor for the relaxation of blood vessels as it came out at the end of a long scientific search. In 1992 NO· was even awarded the title "molecule of the year" by the magazine *Science* (Koshland 1992; Culotta and Koshland 1992). This key and vital activity impressively demonstrates that radical molecules display also central physiological functions. However, as it will be pointed out below, unphysiologically high concentrations of ROS and RNS can directly react with biomolecules, can change their chemical structure and, ultimately, lead to a loss of function. The actual decision in which direction radical function turns is dependent on (1) the concentration, (2) the exact chemical structure, and (3) the site of generation (Halliwell and Gutteridge 1999). Under normal physiological conditions the cell maintains a fine-tuned balance between the generation and removal of ROS and RNS. If the balance of this intracellular redox homeostasis is disturbed radicals accumulate and cause damage. The term *oxidative stress*, first coined by the German biochemist Helmut Sies (Sies 1986), nicely subsumes the negative changes and the damage that may occur as a consequence of the pathophysiological perturbation of the intracellular redox equilibrium.

As shown for the AD-associated amyloid β protein that is found in so-called senile plaques in the brain of Alzheimer patients, oxidative stress can be provoked also by external pathological factors. The approximately 40 amino acid amyloid β peptide interacts with the neuronal membrane and activates membrane-associated NADPH oxidase systems leading to superoxide and subsequent hydrogen peroxide production (Behl et al. 1994). Ultimately, the intracellular peroxides are the substrate for the formation of the highly reactive hydroxyl radicals (OH·) which then oxidize membrane lipids and proteins leading to the destruction of the nerve cell membrane and cell death.

The removal and detoxification of radicals before they accumulate and can do harm is performed by enzymatic and non-enzymatic systems. The water soluble ascorbic acid (vitamin C) and the lipid soluble α-tocopherol (vitamin E) are antioxidants that directly chemically interact with ROS. On the other hand, enzymes such as superoxide dismutase, catalase, and the glutathione/glutathione peroxidase/glutathione reductase-system are of great importance (Moosmann and Behl 2002). As shown in Fig. 3.14 above superoxide can be converted to hydrogen peroxide, a reaction that is carried out by the copper-zinc superoxide dismutase (Cu-Zn-SOD), a key enzyme in the enzymatic oxidative defense. In addition to the Cu-Zn-SOD that is primarily located in the cytoplasm, in mitochondria a manganese-superoxide dismutase (Mn-SOD) exists (Perry et al. 2010). Mutations, in particular, in the copper-zinc-superoxide dismutase are the molecular cause of a set of familial (genetic) forms of a progressive degenerative disorder of motor neurons called amyotrophic lateral sclerosis (ALS; Liscic and Breljak 2011). So, mutations in the genes coding for antioxidant enzymes may lead to lack of or change in enzymatic function which then causes the accumulation of ROS and RNS and oxidative stress; in case of an overboarding accumulation of RNS the disturbance in the redox balance is called nitrosative stress. It should be mentioned here that at least for the ALS cases with a mutation in SOD, the causal relationship between gene mutation/protein function with disease is not that simple. In general, gene mutations causing free radical stress and disease are rare events; of much more interest are the many exogenous factors that induce oxidative and nitrosative stress. The most prominent exogenous conditions in this context are UV light, smoking, chronic bacterial infection or different environmental toxins.

One prominent example of an environmental toxin, the pesticide rotenone, is causing oxidative stress and also the symptoms of Parkinsons disease (PD). The pathogenesis of PD includes oxidative stress after mitochondrial dysfunction, aggregation of the protein α-synuclein, a disturbed protein quality control, excitotoxicity and inflammation. Different labs have shown that by applying rotenone in animal models pathological characteristics typical for PD features are induced. Rotenone is a well-described inhibitor of mitochondrial complex I of the respiratory chain. Interestingly, the rotenone model of PD mimics many clinical symptoms of idiopathic (age-associated sporadic) PD and develops the typical slow but progressive loss of dopaminergic neurons and the Lewy body (protein aggregate) formation in the nigral-striatal system (Xiong et al. 2012). Therefore, in this PD model a direct causal relationship between mitochondrial oxidative stress and neurodegeneration is presented. Consequently, measures that tackle the mitochondrial instability in PD may lead to novel preventive and therapeutic approaches (Moosmann and Behl 2002; Mocko et al. 2010; Jones et al. 2012). So, the interference with the respiratory chain in mitochondria, as caused by the inhibition of complex I with rotenone, may cause an immense flow of free oxygen radicals. And, as clearly shown with the rotenone Parkinson paradigm the induction of oxidative stress can directly cause disease symptoms. But can oxidative stress also directly influence the aging process? At least it has been shown in many investigations that aged tissue has a higher oxidation state, meaning many of the biomolecules found in aged tissue are oxidized and lost their function. Next, a closer look on biomolecules as oxidation targets shall demonstrate the actual consequences on structure and, even more important, on functional changes of the cellular molecules after oxidation by ROS and RNS.

Oxidation of cellular biomolecules, lipids, proteins and nucleic acids: In the membranes of all cells non-saturated fatty acids (**lipids**) are components of the

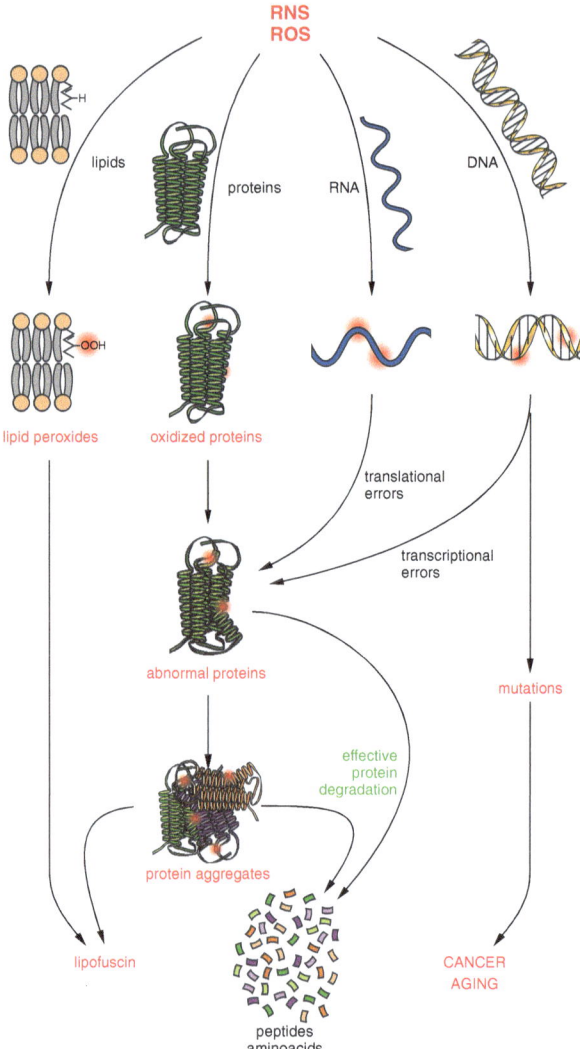

Fig. 3.15 Oxidation products of biomolecules. ROS and RNS chemically react with lipids, proteins and nucleic acids (DNA and RNA) leading to the destruction of the macromolecule or to dysfunction. For instance, oxidized proteins readily form aggregates that can accumulate in cells or oxidatively modified DNA can lead to mutations and, subsequently, to tumor formation

phospholipids that build the membrane double layer. Being non-saturated means that these fatty acids carry one or more double bonds in their hydrophobic tail. These double bonds are ideal targets of oxidation which results in the generation of so called lipid peroxides (see Fig. 3.15). The fatty acid tails of the phospholipids are key structural components for the biophysical features of the membrane double layer and lipid peroxide formation leads to a general hardening of the cellular membrane. Since its fluidity is crucial for the function of membrane-embedded and membrane-attached proteins this structural change may have a great impact on cell function. Even more critical are subsequent chain reactions of lipid peroxides and the disintegration of the membrane that then will be the actual cause of cell death. The AD pathology-associated amyloid β protein has been shown to cause lipid peroxidation in neuronal membranes mediating amyloid β protein-caused oxidative cell death (Behl et al. 1994; Mattson 2009). Taken together, an oxidative attack on the lipids in the membrane may have immediate detrimental effects and the cell membrane breaks up. One should keep in mind that the lipid-protein-bilayer that builds the membrane is the actual boundary of the cell to the extracellular oxidative milieu and is constantly under oxidative attacks from the outside as well as the inside of the cell.

While membrane lipids, and in particular the targets of oxidation, the fatty acids, are rather simple in their chemical structure, **proteins** consisting of amino acids with structurally different side chains are much more complex and the possible reactions of ROS and RNS with amino acid side chains are much more diverse. More than half of the 20 proteinogenic amino acids can react with ROS in a specific manner. Because the integrity of the amino acid side chains is crucial for the forces and interactions that account for the three dimensional protein structure and, subsequently, its correct function, the oxidation of amino acid side chains may change this protein conformation and function significantly. During the course of the reactions of ROS with amino acid side chains even more reactive groups (e.g. aldehyde groups) may be formed which then themselves drive further oxidation reactions with other side chains of the protein or with proximally located proteins leading to actual cross linking. Such cross linking of whole proteins destroys, of course, their function and may, ultimately, lead to the formation of larger protein clusters. High molecular protein clusters are, depending on their size and aggregation status, substrates of protein degradation via proteasome or autophagy. Frequently and especially under constant oxidation conditions, as they may occur during aging, the cell can no longer cope with larger protein aggregates, so that they are deposited inside the cell. Such age-associated deposits, chemically a mixture of oxidized proteins, lipids as well as other components, can be found in cells of the skin, the nerve system, the liver and other organs and are carrying the non-specific name lipofuscin, or age-pigment. The formation of lipofuscin inside cells has been shown to be associated with the life span of post-mitotic cells. The actual amount of accumulated lipofuscin appears to be dependent on the extent of protein oxidation and the effectivity of the cellular protein turnover mechanisms (Jung et al. 2007). And elevated levels of oxidized proteins are found in various age-associated disorders, including atherosclerosis, a panel of neurodegenerative disorders (Alzheimers disease, Parkinsons disease, ALS), and also in cataract, a visible opacity of the eye lense leading to visual loss (Michael and Bron 2011). Inside the cell, oxidized proteins are in large part degraded resulting in the release of the protein constituents, the amino acids that may be used for protein synthesis again. The fact that oxidized proteins do accumulate inside the cell indicates that protein degradation might work only inefficiently (for the description of the two main protein degradation systems, the proteasome and autophagy, see the discussion on proteostasis and aging below). Lipofuscin protein-lipid particles can be observed rather easily employing microscopical methods since they are auto-fluorescing. In aged cell cultures and even in whole organisms, *C. elegans* for instance, fluorescing lipofuscin can be detected and is used as an aging marker (Dunlop et al. 2009). It can be summarized that oxidized proteins are frequently occurring in aging cells. Oxidized proteins become dysfunctional and may escape the degradation mechanisms leading to intracellular protein accumulation, an important hallmark of many age-related disorders. It has to be mentioned here that not only the oxidation of proteins can lead to their intracellular accumulation but rather gene mutations that lead to the formation of proteins with an enhanced aggregation potential, e.g. mutated huntingtin in Huntingtons disease. Coming back to the oxidation of proteins as a result of age-associated oxidative stress, its role for the aging process was summarized best in an landmark review by Earl Stadtman who states that "the importance of protein oxidation in aging is supported by the observation that levels of oxidized proteins increase with animal age. The age-related accumulation of oxidized proteins may reflect age-related increases in rates of ROS generation, decreases in antioxidant activities, or losses in the capacity to degrade oxidized proteins" (Stadtman 2006).

Finally, with respect to protein oxidation it needs to be mentioned that a further characteristic and disease-relevant consequence of oxidative stress in tissue is the so-called glyco-oxidation, which means the formation of sugar-protein-crosslinking products. Glyco-oxidation resulting in the chemical addition of sugar (glucose) onto proteins is irreversible and, ultimately, results in the formation of *advanced glycation endproducts*, AGEs. Most prominently, AGEs are found in Diabetes mellitus patients, where a constantly higher level of glucose leads to its chemical addition to hemoglobin. This oxidized hemoglobin (HbA1c) is increased in untreated Diabetes and is used as a diagnostic serum marker. While HbA1c is an example of a single AGE leading to a dysfunctional hemoglobin and health consequences, most frequently, AGEs that are formed by the oxidation of membrane proteins at the extracellular site may form whole networks leading to a stiffening of tissue. Especially in the extracellular compartment, a tissue hardening in the walls of blood vessels is highly significant and may lead to dysfunction and disease (Gkogkolou and Böhm 2012). In elderly patients with a chronically increased blood glucose level combined with the naturally occurring age-associated oxidative stress, AGEs formation leads to this phenomenon that can be even observed as brownish staining in *post mortem* analysis. It is highly conceivable that a life long challenge of our tissue with ROS and oxidative stress leads to tissue damage in aged organisms.

The main **nucleic acids** in our cells are desoxyribonucleic acid (DNA) and ribonucleic acid (RNA) as the carrier and messenger of genetic information. Mainly the nucleic acid bases contain a number of oxidizable moieties, but ROS may also directly attack the glucose-phosphate backbone of these macromolecules. As with other biomolecules, oxidation of DNA affects its structure and its function, the most prominent changes of the DNA structure being the consequences of oxidation of the nucleic acid bases. Moreover, oxidation of DNA can lead to DNA strand breaks and formation of crosslinks between DNA strands as well as in between the double stranded DNA helix (inter- and intra-strand crosslinks). Furthermore, oxidation reactions may also cause the chemical attachment of proteins to DNA (DNA-protein crosslinks). All such oxidation-based DNA modifications can be the origin of mutations and are observed in a wide range of cancer and various other disorders. A very frequent oxidative DNA-modification is the oxidation of the nucleic acid base desoxyguanosine leading to 8-hydroxy-desoxyguanosine (8-OHdG). Among the bases that are represented in DNA (adenine, guanosine, cytosine, thymine; AGCT) guanosine is least resistant to oxidations (Jena 2012).

As mentioned early in this book the stability of the genome is one of the key issues important for life and the correct structure and function of the DNA is tightly controlled as part of the cell cycle (p53, Rb). Upon DNA damage a cell rather undergoes programmed cell death than continuing cell division and, therefore, carrying on a potentially detrimental mutation. Since the integrity of the DNA is a central concern, different repair mechanisms have evolved that can reverse structural changes and reestablish the correct DNA structure in many cases. If repair fails and the removal of the cell carrying the damaged DNA during cell cycle also fails, damage and/or mutations are transferred from the mother cell to the daughter, eventually leading to cancer development (see chapters on DNA damage and DNA repair above; Alberts et al. 2007; Curtin 2012).

Evolutionary aspects of macromolecule oxidation: Concerning the life span of different species and the extent of oxidative pressure, meaning the accumulation of ROS and oxidative stress, an interesting correlation can be found: in cells of animals with long life spans biochemical constituents are used that are more resistant to oxidations. This applies also to humans especially to tissues with -due to a high oxygen turnover- a rather high mitochondrial ROS production such as heart and brain. Evolution biology and physics tell us that about 2.3 billion years ago the great oxidation event occurred on earth, meaning that the atmosphere started to become enriched in oxygen and aerobic life began to dominate. The ongoing presence of oxygen and the biochemical reactions using oxygen in the respiratory chain to generate energy in form of ATP was and still is a selection pressure and motor of evolution. In fact, the ongoing production of ROS and the resulting oxidative stress is the price aerobic organisms have to pay. At the level of single molecules, this permanent oxidative pressure left its evolutionary traces that can still be found today. It is well described which types of oxidative modifications may occur in reactive amino acid side chains in proteins (Stadtman 2006). When oxidized, some of the amino acids become irrecoverably and irreversibly modified; for others, enzymes exist that reverse the oxidation (e.g. methionine and the methionine sulfoxide reductases). Interestingly, the usage of the in part also irreversibly oxidizable amino acid cysteine is limited in those proteins that are spatially close to the sites of ROS generation in cells. A comparison of the frequency of cysteine occurrence in proteins encoded by the mitochondrial genome of aerobic organisms compared to that in proteins coded for by the nuclear genome revealed an interesting key difference: the amino acid cysteine is extremely deprived in proteins coded by the mitochondrial genome. Other amino acids that are comparable to cysteine in certain parameters, such as methionine and tryptophane and can more easily deal with oxidative attacks, may even be addressed as antioxidatively acting amino acids and are found much

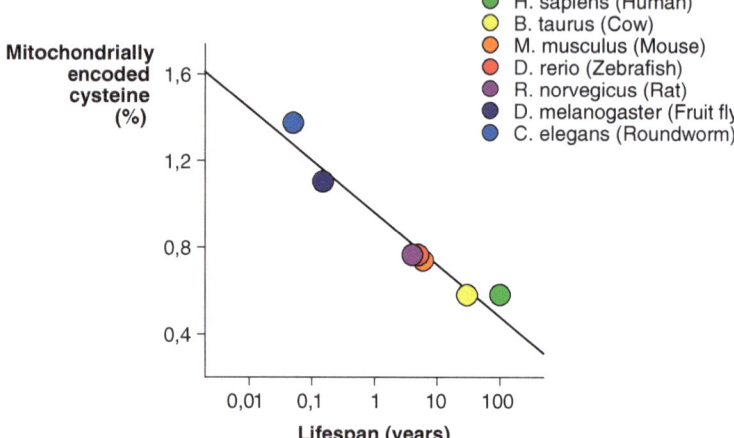

Fig. 3.16 Mitochondrially encoded cysteine and life span. Half-logarithmic linear correlation of maximum life span and mitochondrially encoded protein cysteine content for the species listed (modified after Moosmann and Behl 2008)

more often in mitochondrially coded proteins (Moosmann and Behl 2008). Based on this finding one can argue that the permanent oxidative environment in the mitochondria leaves its traces in the mitochondrial proteins. This could also be addressed as being a sort of structural evolutionary adaptation to oxidative stress in cells. But even more interesting is the observation of a link of the cysteine depletion to life span. When comparing aerobic species from *C. elegans* to *Homo sapiens* it turns out that the extent of depletion in cysteine is directly correlated with the maximum life span of the individual species. The lower the amount of cysteine in proteins coded by the mitochondrial genome is, the higher is the maximum life span (Fig. 3.16; Moosmann and Behl 2008; Schindeldecker et al. 2011). In fact, this correlation is so clear that when focusing on a particular protein (e.g. the core unit of complex IV of the respiratory chain) the maximum life span can be directly estimated based on the number of cysteines. The fact that oxidative stress leaves specific biochemical footprints is strongly supporting the free radical theory of aging. Furthermore, convincing data are presented showing that oxidative stress has also directly shaped the mitochondrial genetic code and there is an adaptive antioxidant methionine accumulation in respiratory chain complexes (Bender et al. 2008). Interestingly, besides these evolutionary aspects and the link to life span, the oxidation of cysteine may even have a direct link to the pathogenesis of age-related disorders such as Parkinsons disease (Meng et al. 2011).

Especially the brain is highly susceptible to oxidative stress and subsequent oxidative damage. This is partly due to the high lipid content of neuronal membranes as a consequence of the function of nerve cells. It is their task to transmit electrical signals over distances (neurotransmission). It is conceivable that this can be achieved much better when the protein content in the cell membrane where the neurotransmission is achieved is lower as compared to other cell types, for instance hepatocytes with different cellular and metabolic functions. In addition, there is a decreased content of antioxidative enzymes and general defense systems in neurons (Halliwell and Gutteridge 1999). Nerve cells when damaged by oxidative stress are dying via apoptosis or necrosis and are ultimately lost. In contrats, cells like erythrocytes have a defined life span of about 4 months; they accumulate oxidative damage but are regularly replaced out of stem cells in the bone marrow via a complex sequence of events called erythropoiesis. Both, nerve cells and erythrocytes do not divide anymore and are post-mitotic differentiated cells. Nerve cells that are lost can not be replaced since there is, compared to other tissues (erythrocytes, mucosa cells, hepatocytes), no significant quantitative regeneration potential. For instance, a brain that experiences a stroke which starts a sequence of pathological events from oxidative stress to inflammation and apoptosis, loses nerve cells. This is also the case for chronic neurodegenerative disorders where oxidative stress is discussed as one cause of nerve cell death (Behl and Moosmann 2002; Dasuri et al. 2013). All these structural and even evolutionary consequences of oxidative stress are highly conceivable. Oxidative modifications change the structure and function of biomolecules leading in most cases to damage and dysfunction. Ongoing oxidative stress is reflected in accumulating tissue damage affecting different organ systems, ultimately precipitating in multimorbidity.

Both, the initial as well as the revisited free radical theory of aging summarizes the accumulating damage to biomolecules as primary cause of aging and age-associated changes (Balaban et al. 2005). When looking through the literature, an increasing number of researchers doubt the correctness of this particular aging theory by referring to novel experimental evidence. Selected examples should be mentioned here: The indirect approach to prove the free radical theory of aging by feeding animals with antioxidants did not show a significant life span extending effect (Strong et al. 2008). Moreover, although contraintuitive the experimental decrease of the enzymatic antioxidant defense system may even lead to a minor life span extension (Ran et al. 2007). It could be that life span extension by antioxidant supplementation is finally proven not to be relevant for higher organisms including humans. Nevertheless, antioxidant supplementation of food, naturally dietary compounds and a specially designed nutrition may be beneficial for various aspects of age-related functional changes and in the end increase-if not life span- metabolic health and the quality of life (Roth et al. 2007; Pan et al. 2012; Horcajada and Offord 2012). Only some nutritional compounds such as flavonoids are acting as antioxidants by directly scavenging free radicals; other constituents of our food intake may act even at the molecular level by modifying gene expression. The genetic and transcriptional effects are investigated by the rather novel research field of *nutrigenomics*. But coming back to the general acceptance of the free radical theory of aging. In general, it is highly criticized that only few studies have actually measured the final outcome of oxidative stress, meaning the determination of direct oxidative damage to tissues, because various experimental systems rather focused on the detection of the ROS generation (for review: Austad 2010). In this very critical view Steven Austad on one side acknowledges that lowering the oxidative load can be beneficial for some tissues and that oxidative stress is implicated in many age-related disorders. But with respect to directly linking aging with free radicals and oxidative stress he proposes: "...as a general explanation for the rate of aging or its modulation there is little point to recommend it" (Austad 2010). This statement is rather strong and, of course, doubts the free radical theory of aging. But as frequently in life the truth may lie somewhere in-between and when comparing the oxidative stress damage in cells with the oxidation of metal (rusting) the free radical theory of aging is rather intriguing, conceivable and easy to follow. But there is one big difference between dead material (e.g. the fender of your old car as target of oxidation) and living cells: cells can in some range react and adapt to external and internal challenges and stresses. Neuronal cells in culture can adapt to oxidants in the culture medium quite easily and amyloid β protein-, glutamate- and hydrogen peroxide-resistant cells have been established showing that life saving adaptations to external challenges are possible.

The free radical theory of aging under stress and ROS as protective signals: Due to the permanent oxidative pressure created internally (mitochondria- and enzyme-derived) and externally (high oxygen intake) throughout the body cells have developed different adaptive strategies including the enhanced expression of antioxidant enzymes. But, in fact, also a major protective role was assigned to free radical generation since ROS affect also gene transcription via the activation of redox-sensitive transcription factors (e.g. NF-kappa B). Recent research efforts go even further in addressing the possibility that ROS show direct beneficial activities by modulating a variety of intracellular processes. Having this in mind the traditional view of the free radical theory of aging suggesting that the aging process is at least in part a consequence of accumulated oxidative damage is challenged. Genetic depletion of the superoxide dismutase (SOD), as the central detoxifying enzyme converting superoxide radicals to hydrogen peroxide in *C. elegans* is not affecting the worms life span. These data as an example and other experimental evidence may indicate that a moderate increase in superoxide-derived oxidative stress leads to an adaptive protective response in the sense of a "prosurvival signaling" as it was stated (Van Raamsdonk et al. 2012). Such findings further strongly challenge the view that the endogenous production of ROS is the simple and sole reason for aging (Liochev 2013). In fact, different cell types can be highly reactive in their adaptive response. In cell culture models it is rather easy to generate cellular subclones that are fully resistant to high concentrations of hydrogen peroxide or oxidative stress inducing neurotoxins such as glutamate by applying these challenges for some time and selecting surviving clones. As early as in 1994 the authors work showed that rat pheocromocytoma cells (derived from a neuroendocrine rat tumor) can adapt to oxidative stress and can develop full resistance to oxidation inducing agents including hydrogen peroxide (Behl et al. 1994). More recently, mouse hippocampal cell clones have been generated that are resistant to hydrogen peroxide and/or glutamate and show a wider range of adaptation including the upregulation of lysosomal degration pathways (Clement et al. 2009, 2010). But even non dividing postmitotic cells in culture can become more resistant to challenges when they survive a first subtoxic hit of the respective toxin, a fact that is easy to conceive when considering the old saying "what does not kill you makes you stronger". In a more physiological context it should be mentioned that adaptive responses were also found in brain areas that are spared from degeneration in Alzheimer patients (Greeve et al. 2000). So, when re-considering the free radical theory of aging on the ground of this collection of novel experimental data and views, one has to acknowledge that it remains a theory that describes potent effectors that modulate the aging process. But these executors of oxidative stress, the ROS as well as the RNS, are not the exclusive inducers of the aging process. The role of free radicals needs to be acknowledged and they can be included as relevant signals that are detrimental or beneficial depending on their actual concentration and the adaptive history of the challenged cells and tissues. Furthermore, the generation rate and metabolic turnover of the individual species is of importance. Concerning the correctness of the free radical theory of aging the pendelum is still moving. Being first extremely on the side of seeing free radicals as central cause of aging it now is swinging over to the other side addressing free radicals as simple byproducts of the age process with no significant influence on the cause. Very likely, oxidative stress is an important part of a more complex network of aging effectors (see Fig. 3.25). The next theory of aging that is rather new may be part of a complex aging network, too. It was already mentioned several times that a cell is investing quite a lot into maintaining the stability of its genome and rather is sent to death than propagating a perhaps detrimental genomic change (mutation). In the next paragraph the role of the maintenance of the intact and functional proteome and the relationship of protein quality control, aging and age-related disease will be addressed. And again, we will encounter oxidative stress.

3.7 Protein Quality Control and Aging

Just as much as genome stability the maintenance and control of the proteome, by definition the entire set of proteins present in a cell at one time, is of importance and proteome changes truly affect the aging process. Proteins are folding to three dimensional functional entities that build the cell structure and act as transporters, receptors and enzymes, including repair enzymes for proteins and the DNA. The folding of a protein into its three dimensional conformation is a highly complex process that is guided and controlled by other proteins acting as helper proteins, so-called chaperones and co-chaperones. Correct folding being a prerequisite for correct function, protein misfolding leads to dysfunction (Fig. 3.17).

The better a species controls its functional entities, the proteins, and the longer the protein function is intact, the longer a life-span could be. This working hypothesis was addressed and partially confirmed by different studies employing naked mole-rats and long-lived bats (Austad 2010). In order to clarify these results at a molecular level the authors lab has started to study the impact of stress (here heat stress) on proteins on the life span of *C. elegans*. Preliminary data indeed confirm that *C. elegans* worms that can cope with heat stress and maintain the protein structure (here of reporter proteins) displays an extended life span compared to *C. elegans* lines where chaperones essential for correct protein folding (and function) are genetically depleted (Kern et al. 2010). But how is protein folding actually accomplished and

Fig. 3.17 Dependence of function on proper folding. Illustrated here for paper boats this holds also true for cellular proteins

controlled, and what happens when proteins are modified, e.g. via oxidation? In the context of this book the question whether protein quality control is altered during aging should be addressed.

A closer (ultrastructural) look into the cytoplasm shows that the cell is a crowded place, full of proteins, nucleic acids, membrane vesicles and organelles, and proteins are always at risk for denaturation. A change in genetic and/or environmental factors and conditions, as it may occur upon extracellular challenges, under pathological conditions and during aging, attacks the integrity of the cellular protein homeostasis (proteostasis). The control of this proteostasis consisting of protein folding, refolding (upon denaturation) and degradation is carried out by a complex network of several hundred evolutionarily highly conserved proteins. Of central importance to this control network are molecular "chaperones", bystander proteins that push the proteins in shape in the course of their synthesis, and the two cellular protein degradation systems, the ubiquitin-proteasome system (UPS) and autophagy. It can be easily imagined that these key players of protein folding and quality control are ubiquitously expressed throughout the organism independently of the tissue type. Indeed, chaperone proteins received great attention in the last decade and are of great interest. To get a closer look into this class of proteins, here, the most prominent chaperone family, heat shock protein 70 (HSP70) family and their regulators should be shortly introduced (Fig. 3.18).

The HSP70 chaperone machinery, UPS and autophagy: The HSP70 proteins (two forms exist, the cognate HSC70 and the inducible form HSP70) and their

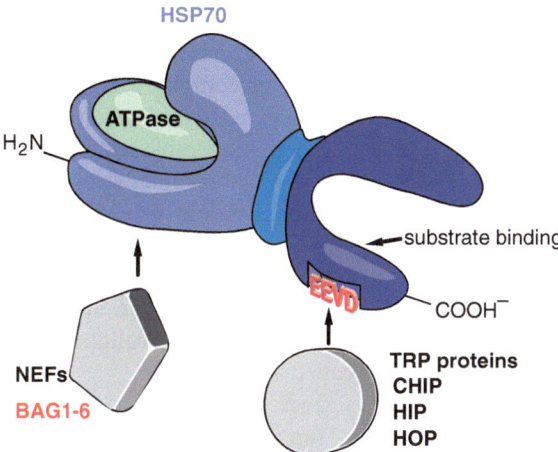

Fig. 3.18 Heat shock protein 70, Hsp70. The chaperone protein HSP70 is interacting with different co-factors and co-chaperones including tetratricopeptide repeat domain (TRP) proteins, the ubiquitin ligase C-terminus of HSC70-Interacting Protein (CHIP), Hsp70-interacting protein (HIP), Hsp70/Hsp90 organizing protein (HOP) and nucleotide exchange factors (NEFs); one group of NEFs are the proteins of the BAG family. Depending on the particular composition of the HSP70 complex protein folding, refolding or degradation is mediated

protein complexes are involved in all processes of proteostasis. First, they play a key role in the correct three dimensional folding of proteins. In the case that proteins are misfolded upon challenges (e.g. heat shock, stress or heavy metal effects) theses chaperone molecules have the capacity of guiding a refolding process. But this refolding activity is limited and if misfolding is too extensive and irreversible proteins are directed to one of the central protein degradation systems, a process that is also controlled by HSP70 (Hartl and Hayer-Hartl 2002; Mayer and Bukau 2005; Bukau et al. 2006; see also Morawe et al. 2012 and references therein). The heat shock protein and chaperone network is tightly controlled itself, mainly through the activity of transcription factors. The heat shock transcription factor 1 (HSF1) is activated by external or internal stress conditions leading to an increased expression of chaperone proteins, enabling the cell to rapidly adjust the cellular chaperone capacity to stress. As already shortly mentioned before, the ubiquitin-proteasome system (UPS) is the major protein degradation machinery of the cell. The UPS is crucial for the regular turnover of proteins since different proteins have different predetermined half-lifes ranging from minutes to months. Therefore, it controls the metabolism and physiological turnover of cytosolic proteins. The name UPS reflects that a degradation-prone protein is modified with the small protein ubiquitin via a specific enzymatic (ubiquitin ligase) process prior to disassembly. This ubiquitinylated protein then is transferred to the multi-enzyme complex of the proteasome. The result of the degradation therein is the cleavage of the amino acid chain down to their basic constituents, the single amino acids, which are then recycled and used for the synthesis of new proteins. In addition to the physiological turnover of proteins the UPS also contributes to the degradation of misfolded proteins (Wilkinson et al. 1980). As effective the UPS as multi-enzyme complex acts, it has also some limitations. The prerequisite for the degradation of proteins via the UPS is the unfolding of the three dimensional structure. But under unfavorable and pathological conditions the amount of misfolded protein can be high, eventually leading to accumulation and then, due to the various non covalent interactions of protein stretches, to the formation of intracellular aggregates. Such high molecular weight protein aggregates cannot be degraded by the UPS. So there is a demand for an additional intracellular pathway, a process that can deal also with larger protein junk. Such an additional pathway for the degradation of proteins and other intracellular debris (e.g. damaged mitochondria) indeed exists and is called autophagy.

The term autophagy summarizes general intracellular degradation pathways that deliver their substrates to lysosomes (Yang and Klionsky 2010) which are intracellular acidified compartments with high proteolytic activity. One major autophagic pathway that is currently in the focus of many research groups worldwide is the so-called macroautophagy. During macroautophagy the degradation substrate is engulfed in a specific membrane structure, the autophagosome. This autophagosome then fuses at one spot with the lysosome thereby forming the autophagolysosome in which the enzymatic degradation actually occurs. Macroautophagy can be executed rather unspecifically with respect to the degradation substrate but can occur also in a selective manner, then called selective macroautophagy (Dikic et al. 2010; Wong et al. 2011; Gamerdinger et al. 2011a). In selective macroautophagy targeting aggregates proteins, HSP70 again plays a role since firstly the degradation substrate is recognized by and binds to HSP70. Further steps lead to the assembly of the degradation-prone proteins at a certain (frequently perinuclear) site in the cell. The transfer of the degradation substrates into the membrane-driven autophagy process can be mediated, for instance, by the co-chaperone BAG3 (BCL2-associated athanogene; name 'athanogene' from the Greek word *athánatos*, meaning 'against death') and then is called *BAG3-mediated selective macroautophagy* (Gamerdinger et al. 2009). Taken together: when encountering a misbalance of the proteostasis network resulting in an increased demand for protein degradation the cell has two principal options to degrade target proteins via (1) the proteasome and (2) autophagy. During cell aging as well as a consequence of various disorders the proteostasis can be acutely or chronically challenged and the degradation machineries are turned on to prevent an overboarding accumulation of protein aggregates (David et al. 2010; Morawe et al. 2012) (Fig. 3.19).

Protein quality control and proteostasis in aged cells and aged organisms: Chaperones (e.g. HSP70) and co-chaperones (e.g. BAG proteins) are crucial for the successful performance of intracellular protein degradation. It is of great interest to study these pathways in the context of aging and disease as for a number of age-associated neurodegenerative disorders protein aggregates are well accepted pathological hallmarks (Soto 2008). Moreover, it is necessary to uncover the molecular details of a possible tissue-specific chaperone activity since different tissues may have quite different requirements for protein quality control (e.g. nerve cells vs.

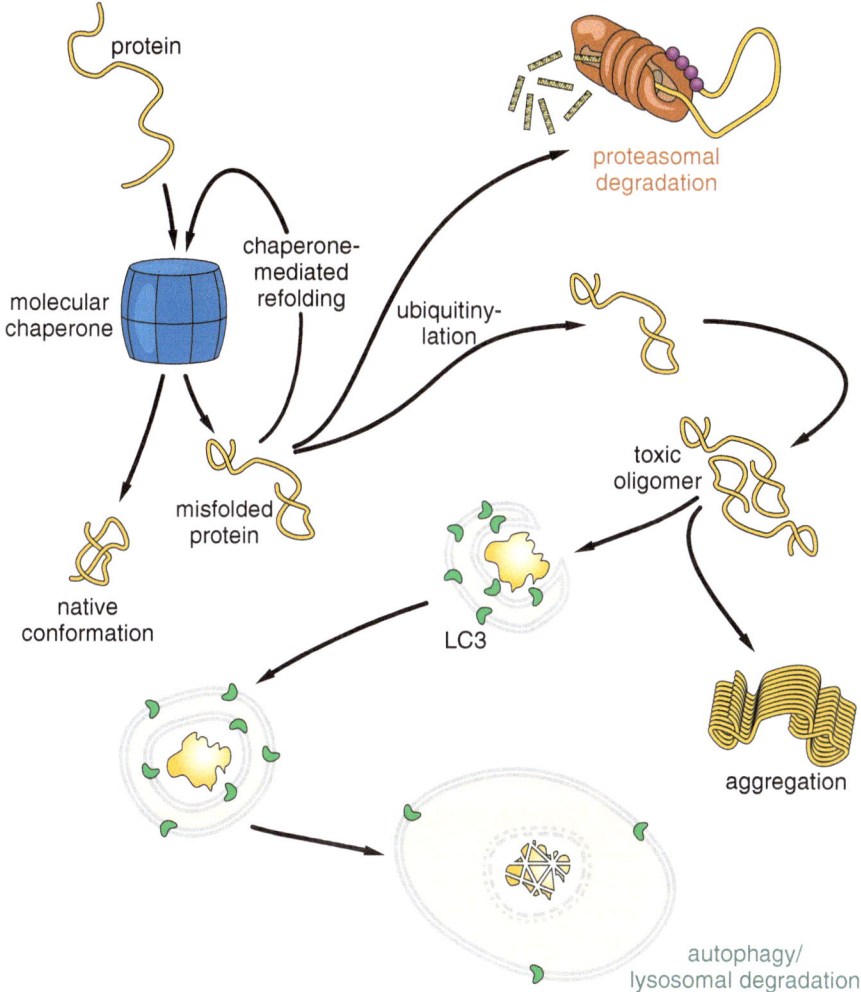

Fig. 3.19 Proteasomal and autophagic degradation, two principal pathways of protein decomposition. Cellular proteins have a certain half life and need to be degraded which is usually carried out by a multi-enzyme complex, the proteasome. Misfolded proteins can be broken down by proteasomal degradation as well usually becoming ubiqitinated before. Larger conglomerates (toxic oligomers, high molecular weight aggregates) can no longer be handled by the proteasome but are degraded via the lysosomal autophagy pathway

blood cells). The authors lab analyzed the chaperone activity and capacity in *C. elegans* by comparing neuronal and muscular tissue. And, indeed, these tissues display different protein folding and refolding activities when the worms are subjected to heat stress. It was found that compared to muscle cells neurons are particularly sensitive to protein denaturation during heat stress (Kern et al. 2010). But what is also of great importance is a possible adaptation of the chaperone network during the progress of aging since protein aggregation and disruption of proteostasis are characteristic for aged cells (David et al. 2010). Several studies employing model systems of aging research have clearly demonstrated a role of chaperones in aging of an organism. When *Drosophila melanogaster* or *C. elegans* were set under a repetitive mild heat stress, the mortality of both organisms was decreased, a phenomenon that is mediated via HSP70. And when the level of the chaperone HSP70 is experimentally increased the life span of the worm rises, too (Tatar et al. 1997). The levels of the mRNA and protein of this chaperone in aged cells were found to be either at basal amounts or increased but in any case the heat stress-mediated induction of chaperone expression and the control of the proteostasis was impaired in aged cells (e.g. Kaarniranta et al. 2009). Obviously, the basic chaperone machinery is still present in aged cells but the reactivity to external stress is diminished. As already mentioned above, upon heat stress the transcription of chaperone genes is controlled by the transcription factor HSF1. HSF1 has been found to exert an impaired DNA-binding potential in aged cells explaining the diminished heat stress response in aged cells (Heydari et al. 2000). Consistently, a lower HSF1 activity shortens the life span of *C. elegans* and an increased expression prolongs life. Finally, it is known that the activity of HSF1 and thereby presumably of the HSF1-controlled chaperone network of proteins is essential for the extended life span of the extremely long-lived *C. elegans* worm lines daf-2/Insulin/IGF-1 receptor mutants that were already discussed in the context of caloric restriction and life span (Hsu et al. 2003; see also respective chapter above). Taken together, the transcriptional effects of HSF1 and the subsequently induced chaperone activity can directly promote longevity of model organisms.

Obviously, when organisms age, protein quality control and the regulation of protein homeostasis become disrupted and challenge the cells to adapt to the new "protein stress" situation. As outlined in an elegant recent review article several alterations in the principal protein degradation pathways and mechanisms can happen during aging (David 2012). It has been observed that in mammals the so-called unfolded protein response (UPR) activated by ER stress is impaired with age (Brown and Naidoo 2012). The UPR has important cellular functions and is provoked by unfolded or misfolded proteins that occur in the lumen of the endoplasmic reticulum (ER), the compartment that carries the secretory proteins and mediates their secretion into the extracellular compartment. Basically, when defect proteins appear as cause of ER stress, it is one task of the UPR to restore normal cellular function by halting further protein translation. In addition, the UPR activates signaling cascades leading to an increased production of molecular chaperones ensuring proper protein folding and refolding. In case the UPR is not able to reverse the un- or misfolding of the proteins in the ER, the UPR is linked also to programmed cell death (apoptosis). Moreover, age-associated changes in proteasomal activity have been found. As observed in rats, proteasomal activity is decreasing in certain tissues with age (Anselmi et al. 1998). In addition, the lysosomal chaperone-mediated autophagy (CMA) activity has been reported to be reduced in the liver of aged rats as well as in human senescent fibroblasts (Cuervo and Dice 2000). Interestingly, oxidative stress also plays a role in the context of protein degradation: oxidations and nitrations of proteins in the course of cellular aging lead to an impaired protein degradation (Squier 2001;

Poon 2006). Mutations and age-related changes in mRNA transcription and protein translation may lead to defective proteins that challenge the protein quality control system, eventually causing protein aggregation disorders, including Huntingtons, Alzheimers and Parkinsons disease (van Leeuwen et al. 1998; de Pril et al. 2004; Morawe et al. 2012). So the changes in protein biochemistry during aging are quite a few and all these pathobiochemical alterations may contribute to a widespread altered proteostasis leading to protein misfolding, dysfunction and aggregation. In summary, all these observations undoubtedly demonstrate a clear association of chaperones, proteostasis, protein quality control and aging.

Shifting gears in protein degradation during aging, the BAG1/BAG3-switch: There is recent evidence that during aging of cells and whole organisms the principal pathways of protein degradation are directly modulated. Here, work of the authors group should be shortly introduced: Employing an experimental approach analyzing basic biochemical differences of young and old (non-transformed) human cells, the expression of two members of a particular family of co-chaperone proteins, the BAG family members BAG1 and BAG3, was identified as being reciprocally regulated during the cellular aging process (Gamerdinger et al. 2009). This reciprocal expression changes we called the age-related "BAG1/BAG3-switch". Under normal physiological conditions HSP70 recognizes misfolded proteins and links them to the UPS assisted by the co-chaperone BAG1. When pathophysiological conditions associated with cellular aging or with acute (e.g. oxidative) stress occur that are leading to an accumulation of misfolded and aggregated proteins the proteasomeal degradation system is no longer able to cope with the increased protein degradation demand. The selective macroautophagy pathway turned on then was named *BAG3-mediated selective macroautophagy* (Gamerdinger et al. 2009) since the co-chaperone BAG3 is essential for it (Fig. 3.20).

With respect to its particular protein domain structure the co-chaperone BAG3 is a highly promiscuous molecule and can interact with different proteins linking it to several intracellular pathways including apoptosis, differentiation and adhesion (McCollum et al. 2010). Besides the regulatory function of the two BAG-proteins and the BAG1/BAG3-switch in the course of aging the key to the degradation process is the specificity of HSP70 chaperones to misfolded proteins. The HSP70 chaperone system as a multifunctional complex of proteins depends in its function on the particular interaction partners (Hartl and Hayer-Hartl 2002). Thereby, HSP70 is the core chaperone and many co-chaperones and co-regulators may physically interact. Interestingly, the individual binding of substrates to HSP70 and the subsequent necessary release of the substrate from the HSP70 protein complex is controlled by the cellular energy currency ATP. Protein folding and refolding activities of HSP70 as well as protein degradation activity linking it to the proteasome or to macroautophagy are determined by the association of HSP70 with particular co-chaperones such as BAG1 and BAG3 (as nucleotide exchange factors) that directly influence the actual turnover of ATP at HSP70 (Young 2010). How exactly HSP70 directs degradation-prone and aggregated proteins to the particular pathway is summarized elsewhere (Gamerdinger et al. 2011b). Taken together, the identification of this novel pathway and its induction during the cellular aging process -when the demand for

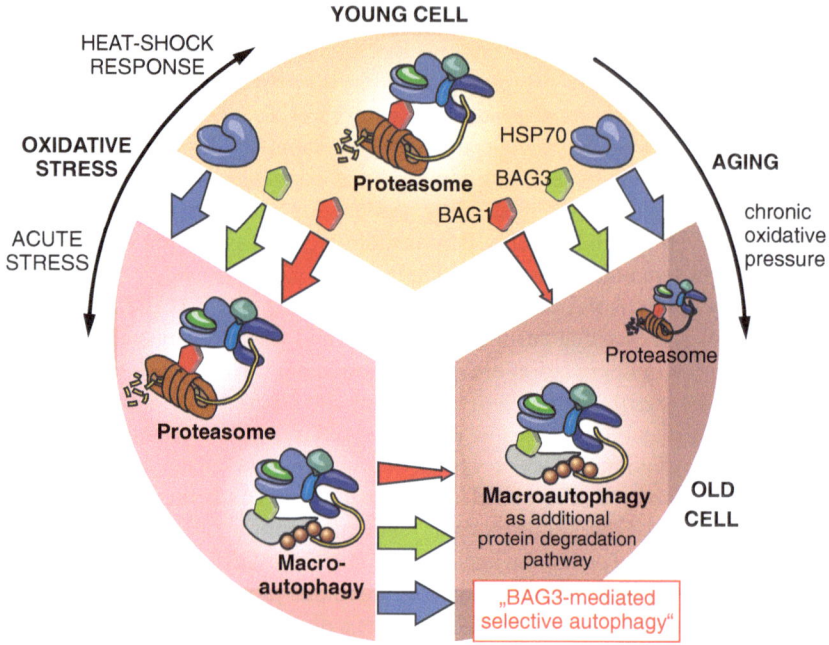

Fig. 3.20 The BAG1/BAG3-switch and "BAG3-mediated selective autophagy". In young cells mainly the proteasome is active and autophagic activity is comparably low; BAG1 is essential for the transfer of substrates to the proteasome. During aging (or upon acute oxidative stress) the expression of BAG3 increases and in addition to the proteasomal degradation selective autophagy depending on BAG3 is induced. This change in BAG protein expression is called BAG1/BAG3-switch and degradation-prone proteins in aged cells are degraded by this "BAG3-mediated selective autophagy" (taken from Gamerdinger et al. 2011b)

effective protein degradation is increased- showed that the protein quality control system is highly adaptive. Obviously, intracellular pathways are highly sensitive to age-associated conditions. Protein misfolding and aggregation are frequent companions of aging and various neurodegenerative disorders. So this adaptation may serve as a response also to disease conditions based on the disturbance of the intracellular protein homeostasis. As mentioned, an enhanced protein aggregation is a key hallmark of aging and in particular of various neurodegenerative disorders in humans that are still without an effective causal therapy, such as ALS, Alzheimers, Huntingtons and Parkinsons disease (Morawe et al. 2012). The BAG3-mediated selective macroautophagy pathway has been shown to effectively cope with protein aggregates, such as those occurring when SOD is mutated (Gamerdinger et al. 2011a). But in age-associated neurodegenerative disorders, such as ALS characterized by intracellular SOD aggregates, the protein degradation, obviously, is overruled by the massive protein aggregation that occurs. What one may learn from the *BAG1/BAG3-switch* actually is that there is an adaptation capacity during cellular aging that needs to be considered when investigating the cause and potential novel therapies of so far untreatable age-associated neurodegenerative diseases. One approach that would be too good to be true if it would really work is the stabilization or stimulation of the intracellular protein quality control system to stabilize neurons during the aging process so that upcoming protein stress can be prevented before it causes harm. This view for sure is an oversimplification with respect to the exact cause of the mentioned age-associated disorders since for the large number of mutation-independent and sporadic (strictly age-related) cases of neurodegeneration it is not really clear whether protein accumulation is actually *the* cause or may only be a byproduct of an otherwise initiated and ongoing disease process. On the other hand, knowing that nerve cells that undergo degeneration in disease (cortical neurons in AD, dopaminergic neurons in PD, motor neurons in ALS) are all post-mitotic for life time and, consequently, face multiple (environmental and other) challenges during life and along with aging, it could be a principal strategy to stabilize key processes of intracellular physiology such as the protein quality control and especially the protein degradation pathways to make nerve cells more resistant to age-related and potentially pathogenic conditions. Therefore, a specific stabilization and age-associated induction of the BAG3-mediated selective macroautophagy pathway could be one way to increase survival chances of a neuron that faces an age- and/or disease-related disturbance of protein homeostasis.

Interestingly, the closer look into cellular processes of aging always directly leads to the topic of cause and consequence of disorders that accompany aging and increase in number. The authors view is that a full comprehension of neurodegenerative disorders including also AD, which is one huge burden of our aging society, and the development of causal therapies is dependent on decoding cell aging and age-associated biochemistry. So understanding aging of neurons is the ultimate prerequisite of uncovering the cause/s of so far still detrimental neurodegenerative disorders of the high age. As simple as this may sound, as surprising it is that such a view has been almost neglected by scientists for many years, partially driven by the approach that the discovery of mutations specific for genetic/familial forms of disease may almost instantly lead to the understanding also of the non-genetic cases which are usually the majority. A very good example definitively is the AD research of the last three decades with the outcome that we know almost all details on the cause of the very rare cases of familial AD and still have no cure (or at least a promising pharmacological targeting concept) for the sporadic non-familial AD cases that are much more relevant due to their constantly increasing number.

In the next chapter the focus of the discussion will be shifted back again to processes, changes and hallmarks that are directly associated with the physiological aging process. And here, the attention will be drawn back to the nucleus and the genomic DNA.

3.8 Epigenetics

The importance of a stable genome and the cells investment for the prevention of mutations leading to dysfunction and, eventually, disease has already been emphasized: genome stability is mandatory for life and inheritance. In modern molecular medical research an intensive search is going on for disease-causing and disease-associated single genes, gene families or gene patterns. Human genetics and bioinformatics have developed smart toolboxes to screen whole genomes and follow family trees and gene traits with the goal to identify such disease-relevant genes. These genes and the derived proteins are pharmacological targets for prevention and intervention. In similar human genetic approaches the search is ongoing for genes that are associated with an extended life span collecting and subsequently analyzing, for instance, whole populations of centenarians. While the hope is to uncover the genetic signature of exceptional longevity, so far no single key anti-aging gene has been found in humans but always rather complex patterns of whole sets of genes associated with extended life spans (Tan et al. 2009; Sebastiani et al. 2012). As already mentioned in this book and while collecting and reviewing theories and mechanisms of aging one can conclude that it is very unlikely that *the* ultimate central regulator gene of aging exists. On the other hand, the individual differences in life span observed within one species seem to be affected by extrinsic environmental factors. Such factors can slightly modify the chemical structure of components of the DNA. In fact, while the human genome is highly stable concerning its general structure and the sequence of the DNA, chemically its components can be constantly modified leading to alterations in gene expression patterns and, at the level of the whole organism, to changes in function. These chemical modifications of our genome that occur on top ("epi-") of the DNA sequence are investigated in the research field of *epigenetics*. A recent consensus definition states: "An epigenetic trait is a stable heritable phenotype resulting from changes in a chromosome without alterations in the DNA sequence" (Berger et al. 2009).

Small changes in chemistry, great effects in gene regulation-the molecular correlates of epigenetics: The research field of epigenetics currently appears as one of the directions that may uncover new mechanisms of development and disease and may strengthen our understanding on the heritage of genetic predisposition for certain pathological conditions. For humans the study of identical twins are of special value (Steves et al. 2012). Worldwide epigenetic research is intensified and the more progress is made it appears that epigenetic mechanisms are involved in all key functions of a cell. Therefore, they are reflected in consequences for physiology and pathophysiology as well as for aging. As frequently in science, the levels of complexity are changing along with the increasing knowledge on certain functions of interest. It seems, for instance, in the search for the cause of detrimental human disorders including Alzheimers disease, there is always one more level of upstream or downstream key regulation that needs to be investigated before the actual molecular cause of the onset of complex age-associated pathologies can be fixed. And indeed, epigenetic changes have been shown to be highly associated with a wide range of human disorders (Adwan and Zawia 2013; Kaelin and McKnight 2013; Gonzalo 2010; D'Aquila et al. 2013). Interestingly, the term epigenetics has been coined a long time ago. One of the key scientists here was Conrad Hal Waddington (1905–1975), a developmental biologist, geneticist, and embryologist, actually, he was a pioneer of systems biology. In an early publication Waddington already used the term "epigenetic landscape" which at that time for the research community was a metaphor for the modulation of development by gene regulation as recently re-published in a *reprint-and-reflection-article* to honor the visions of this early epigeneticist (Waddington 2012). Today, the epigenetic landscape is more than ever in the focus also of aging research. But before presenting some of the links of epigenetics and aging, here the molecular substrate, marks and types of epigenetic changes in the genome should be shortly described.

In general, epigenetics addresses all hereditable, i.e. mitosis- (and sometimes also meiosis-) transmitted alterations of a cellular phenotype. Besides the differential expression of transcription factors, these changes are responsible for a cell-type specific gene expression but do not affect the DNA base pairs and their sequence. To make it more clear: all cells of an organism contain the same nuclear genome, i.e. the DNA sequence is identical, but the phenotype and function of, for instance, nerve cells (e.g. ischiatic neurons, pyramidal neurons in the hippocampus) are quite different from other cells in the body (e.g. skin cells, hepatocytes, immune cells). These differences are maintained throughout the life span of an individual cell guaranteeing this differential phenotype (leading to diverging functions) and are caused by the differential expression of genes occurring in the various cell types. Certain genes are active in one tissue but not in others. This silencing of the transcription of a large array of genes is made possible by slight chemical modifications of the genome. Epigenetic marks allow such differential expression patterns. The chemistry behind it is partly rather simple and concerns (1) the methylation of the DNA, (2) different modifications of the nuclear histone proteins, and (3) to some extent also the expression of non-coding RNAs (ncRNAs). The so far best-understood epigenetic marks are DNA methylation and post-translational modifications of histones (Fig. 3.21).

Methylation of the DNA and links to the aging process: Methylation of the DNA means the covalent addition of a methyl group ($-CH_3$) to the DNA base cytosine (resulting in 5-methylcytosine) that in most cases is located near the base guanosine. These hotspots of DNA methylation are called "CpG islands". Various biological functions during development and differentiation are regulated via DNA methylation. Most impressively: also the inactivation of one of the two X-chromosomes in females is mediated by this epigenetic change. The biochemical machinery performing the methylation consists of various DNA methyltransferases (DNMTs). The addition of the small CH_3-group leads to ultrastructural rearrangements preventing the transcription of the methylated gene stretches via various mechanisms (He et al. 2011). In fact, the methylation leads to a close chromatin state through the attraction of protein complexes with repressive activities, ultimately leading to switching off genes. Taken together, DNA methylation alters gene expression in dividing cells on their way of differentiation from embryonic stem cells to specialized tissue. Thereby, it acts globally as well as on specific gene loci. Some time ago the authors lab has

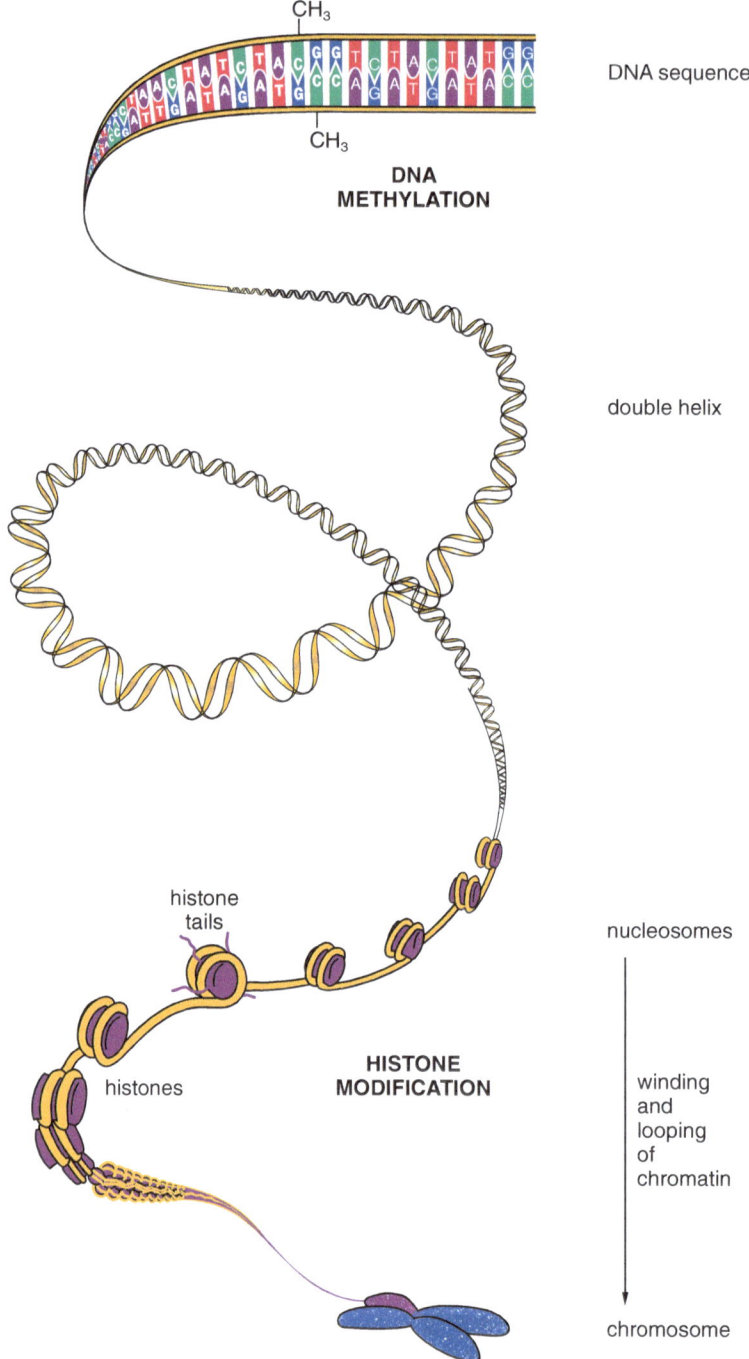

Fig. 3.21 Main components of the epigenetic code. The biochemical hallmarks of epigenetics are DNA methylation and histone modification (modified after Qiu 2006)

demonstrated that DNA methylation mediated by estrogen receptors can lead to an almost complete silencing of selected genes. In a cellular study employing human neuroblastoma cells, the presence of a particular subtype of the human estrogen receptor (estrogen receptor alpha) suppressed the expression of the proteins caveolin 1 and 2 which are involved in endocytosis and intracellular signaling. A detailed molecular analysis revealed that the observed decreased caveolin expression was accompanied by changes in the methylation pattern of the caveolin promoters. In selected promoter regions of the human caveolin-1 gene certain CpG dinucleotides were hypermethylated in the human neuroblastoma cells expressing the estrogen receptor alpha. On the other hand, the same DNA sites were unmethylated in control cells with no estrogen receptors or cells that expressed the other subtype (estrogen receptor beta). An in vivo analysis detected a down-regulation of caveolin-1 expression also after long term estrogen exposure in certain regions of the mouse brain (Zschocke et al. 2002). In experimental setups one may interfere with and even prevent the DNA methylation process. Physiologically, key epigenetic alterations occurring early in the development of a cell are usually of permanent nature and prevent its reverse development back into a stem cell or the differentiation into a completely different cell type. DNA methylation modifications also mediate a process called *genomic imprinting* that ensures the expression of only one allele from either the maternal or paternal source.

Early and despite their age still acknowledged hallmark investigations revealed an overall loss of DNA methylation in rat brain, and heart and also in other species and tissues, during aging (Vanyushin et al. 1973a, b; Romanov and Vanyushin 1981). Interestingly, an age-associated reduction in the methylation of the DNA of the inactive X-chromosome has been observed (Busque et sl. 1996). On the other hand, as recently reviewed, the total number of altered methylation sites may also increase with higher age and is, therefore, proposed as a marker for chronological age (Ben-Avraham et al. 2012). In fact, a whole range of specific gene loci show an increased DNA methylation status, they become hypermethylated, in normal tissues during aging (Fraga and Esteller 2007), including the estrogen receptor, insulin-like growth factor II, the transcription factor c-fos and others (Gonzalo 2010). The epigenetic changes (the *epigenome*) can be altered through the influence of genetic, environmental and stochastic factors. In support of this, an age-dependent change in the overall level of DNA methylation has been observed, for instance, in a human population from Iceland and Utah showing also a familial clustering of DNA methylation (Bjornsson et al. 2008). Most importantly, exogenous factors can affect the methylation status over time. Dietary deficiency in certain nutrients but also UV light has a direct influence on the methylation status depending on the tissue and age are examples (e.g. Chouliaras et al. 2012). Taken all these studies together, it appears that during aging the genome experiences a global *hypo*methylation while single gene promoters are undergoing a specific *hyper*methylation. So far, all these findings are rather descriptive and the exact molecular mechanisms behind these global down- and individual gene-specific up-regulation of DNA methylation during aging are not yet clear. There is no doubt, already when focusing only on one particular epigenetic mark, the DNA methylation, it appears that during aging many changes occur. But before formulating a strong *epigenetic theory of aging*, possible causal and functional links between these chemical modifications and the aging process need to be undoubtedly uncovered. This also applies to the epigenetic changes which will be discussed next, the different post-translational modifications of the histones.

Histone modifications and links to aging: Histones are a family of proteins in the eukaryotic cell nucleus. Due to their highly alkaline character -meaning that they carry stretches of the positively charged amino acids lysine and arginine- histones enable the packaging of the DNA into structural units called *nucleosomes*. By definition a nucleosome consists of 147 base pairs of double-stranded DNA that is wound around an octamer consisting of two H2A, H2B, H3, H4 proteins each (Fig. 3.22). An additional histone (H1) then links the nucleosomes and, therefore, helps to line up the DNA in the high-ordered structure of chromatin. Unwound DNA would be very long and so these structural measures to tightly pack the chromosomal DNA thread are essential (see Alberts et al. 2007). As this nuclear proteins have a key structural role in DNA packing, histones also modulate the regulation of gene expression. Proteins in general can be chemically modified after translation, which is a frequently occurring process and most prominently exemplified by phosphorylation, glycosylation, acetylation and methylation of specific amino acid side chains. In fact, post-translational modifications are of great importance for protein function (e.g. enzymes often are activated by the addition of inorganic phosphate groups; see also chapter on cell cycle control). The addition of a chemical group to certain amino acid side chains alters their chemical and, therefore, biophysical properties. The knowledge that the three dimensional conformation of a protein is highly depending on the covalent and, even more important, the non-covalent interactions of the amino acid side chains, explains that the addition of a novel chemical moiety alters the binding forces, the protein structure and, ultimately, the protein function. All this applies to histone proteins that are targets of post-translational modifications. Since the major role of histones is packing of the DNA, these chemical alterations are epigenetic modifications and influence gene function without altering the DNA sequence.

Histones potentially undergo the following chemical modifications: acetylation (addition of an acetyl group), methylation (addition of a methyl group), phosphorylation (addition of an inorganic phosphate), and ADP ribosylation (addition of one or more ADP-ribose moieties) to name the most frequent ones. These changes can occur at the amino- and carboxy-termini as well as at the core domain of the histone molecule and alter the organization of the affected nucleosome leading to DNA that is transcriptionally active or completely inactive (silent), depending on the specific type of modification. Moreover, chemically modified histones also differently interact with other transcriptionally relevant molecules such as transcription factors. Given the importance of the particular chemical modification status of histones involved in nucleosome formation, even the term *histone code* has been coined, underlining the impact of such modifications on the transcriptional process.

Besides DNA methylation the most prominent epigenetic marks are histone acetylation and histone methylation. The transfer of an acetyl group is carried out by histone acetyl transferases (HATs; Marmorstein and Roth 2001; Zentner and Henikoff 2013) which comprises a whole enzyme superfamily. Due to their key functions

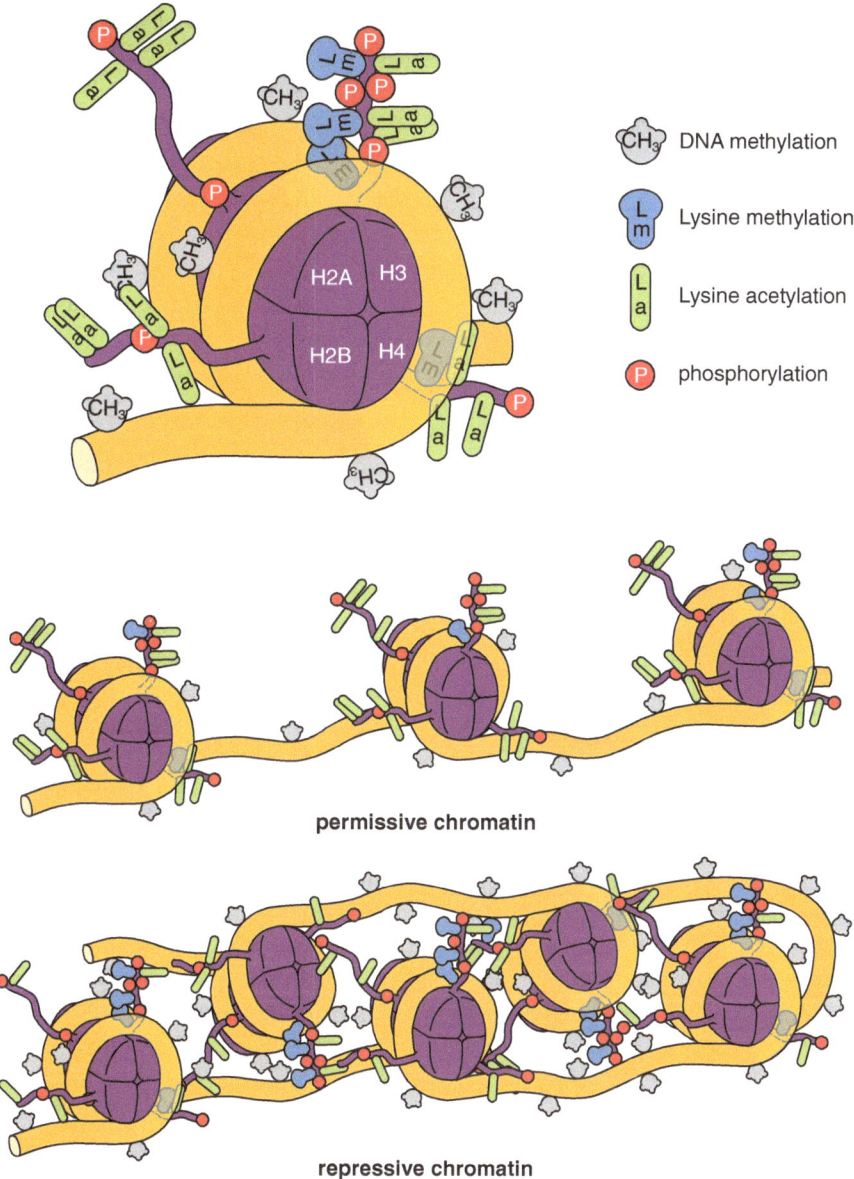

Fig. 3.22 Histone modification and chromatin structure. Schematic representation of a nucleosome: the DNA double strand wrapped around an octamer of histone molecules H2A, H2B, H3 and H4 each of which can be posttranslationally modified. Permissive, i.e. transcriptionally accessible, chromatin is characterized by hyperacetylation of histone tails and hypomethylation of histones and DNA. Repressive since more condensed chromatin exhibits hypoacetylation of histones and hypermethylation of histones and DNA (modified after Gonzalo 2010)

in transcription the different HATs are linked to various diseases including cancer (Barneda-Zahonero and Parra 2012; Pirooznia 2013). Mechanistically, the effect of acetylation (and deacetylation) on the transcriptional activity is highly comprehensible. Shortly: the transfer of an acetyl-group to positively charged amino acid side chains (mainly lysine) in the histone molecule masks the positive charge of the amino acid that is crucial for the non-covalent binding to the overall negatively charged DNA. A tight binding of non acetylated lysines of the histones to the DNA allows tight condensation of the histone-DNA complex, the acetylation loosens this interaction resulting in chromatin decondensation (looser packaging). Transcription factors can now access the DNA. The reversal of this reaction meaning the removal of the acetyl group from the modified amino acids (deacetylation) is performed by histone deacetylases (HDACs) leading to the reestablishment of the tight histone-DNA interaction. While upon acetylation transcription is promoted, following deacetylation transcription is repressed (silent). Therefore, acetylation is frequently occurring at those sites in histones that are associated with the gene promoter region on the DNA underlining the switch from a silent transcriptional status (*repressive chromatin*) to an active one (*permissive chromatin*) in the acetylation/deacetylation cycle (Marmorstein and Roth 2001; Zentner and Henikoff 2013).

The positive charges of lysine and arginine side chains in the histones can also be altered via the addition of a methyl group (methylation) and this modification plays an important role in the regulation of transcription, too. The transfer of a methyl group to the amino acids is also performed by specific enzymes, the histone methyltransferases (HMTs), and methylation can occur to a different extent. Interestingly, it is of key importance whether the targeted lysine undergoes mono-, di- or trimethylation, since the grade of methylation determines the degree of chromatin condensation leading to differential effects on gene transcription inhibition (Gonzalo 2010). All potential histone modifications (differential acetylation and methylation) are key epigentic control factors and present the chromatin as a highly flexible structure with respect to the transciptional status. And there are links of histone modifications to aging.

In experimental aging models (in vitro and in vivo) an overall increase of the methylation status of certain histone subtypes is observed (Bártová et al. 2008; Ben-Avraham et al. 2012). In mouse models changes in histone acetylation could be directly linked to an age-related decrease in cognitive function (Greer et al. 2010). Sirtuin 6 (SIRT6) is a deacetylase enzyme and involved in telomere function and in the expression of age-associated genes. Its loss leads to a progeria phenotype of premature aging (Peleg et al. 2010). The genetic deletion of Sirt6 in mice results in a severe degenerative phenotype with impaired liver function and premature death (Marquardt et al. 2013). Significant changes in the pattern of histone modifications are also found in different progeria syndromes such as Hutchinson-Gilford Progeria but also in aged human tissue in general (McCord et al. 2009). An extensive histone methylation (hypermethylation) at sites linked to the retinoblastoma tumor suppressor gene (but demethylation as well) may affect retinoblastoma target genes in aged cells (Shumaker et al. 2006). Moreover, the "old companions" showing up in different contexts in the presented discussion, the insulin/IGF-1 genes, can be modulated in their expression via histone methylation (D'Aquila et al. 2013). Finally, the link between aging and several histone-modifying enzymes should be mentioned. Here, the family of sirtuins should be highlighted which, as already mentioned before, comprises a whole group of evolutionary highly conserved nicotinamide adenine dinucleotide- (NAD-) dependent enzymes that regulate life span in many model organisms including yeast and mice (Burgess et al. 2010). During aging a decline in SIR2 protein has been observed in cellular model systems resulting in an increased histone acetylation (McGuinness et al. 2011), thereby affecting gene transcription. Although different studies on sirtuins and their direct link to aging are currently discussed quite controversially (see chapter on sirtuins), a connection of the activity of sirtuins to oxidative stress, shown to be involved in aging as well, was presented and a conceptual model putting the sirtuins on the map of a ROS-driven mitochondria-mediated hormetic (meaning protective) response is currently put forward (Merksamer et al. 2013).

Non-coding RNAs as novel players in epigenetics and aging: There are several types of RNA molecules present in the cell. The most intensively studied ones are messenger RNAs (mRNAs) that are direct transcripts of coding genes and the templates for protein translation, and the transfer RNAs (tRNAs) that carry specific amino acids to the protein translation process at the ribosome. In addition, also other non coding RNAs (ncRNAs) exist of which there is still only a fragmentary understanding concerning their function. Constitutively expressed forms of ncRNAs as well as regulatory ncRNAs are known and among the latter the so-called micro RNAs (miRNAs) and small interfering RNAs (siRNAs) are functionally characterized to some extent. The miRNAs negatively regulate gene expression by directly targeting the cellular mRNAs. In mammals, single miRNA species can target many different mRNAs which can lead to significant changes in a wide range of gene expression. Micro RNAs are encoded by the nuclear DNA and act by base-pairing with complementary sequences within the mRNA molecules. The consequence is the suppression of protein expression either via the repression of the translation of the respective mRNA (if base-pairing is imperfect) or via the induction of the degradation of its corresponding mRNA (if base-pairing is perfect) (Fig. 3.23).

It is estimated that the human genome encodes over a thousand different miRNAs that potentially target approximately 60 % of genes. The fact that miRNAs are highly conserved throughout eukaryotic organisms suggests that these molecules may represent early evolutionary regulators of gene transcription. Again, first evidence of an involvement of miRNA in aging and life span came from *C. elegans* studies. A miRNA named lin-4 has been shown to affect the life span of the worm, actually by regulating the lin-14 mRNA and thus the insulin signaling pathway (Lee et al. 1993). Recently, technology platforms have been developed that provide a methodological approach to uncover aging-related miRNAs and their targets. In fact, so far miRNAs are investigated at best for their influence on aging and life span. Different miRNAs besides lin-4 (e.g. miR1, miR-145, miR-140) target the insulin/IGF-1 receptor and related signaling molecules. And obviously again, the insulin/IGF pathway is a molecular downstream target of regulatory molecules that influence aging (Ibáñez-Ventoso and Driscoll 2009; Grillari and Grillari-Voglauer 2010; Jung and Suh 2012). When looking through the literature additional evidence for an important

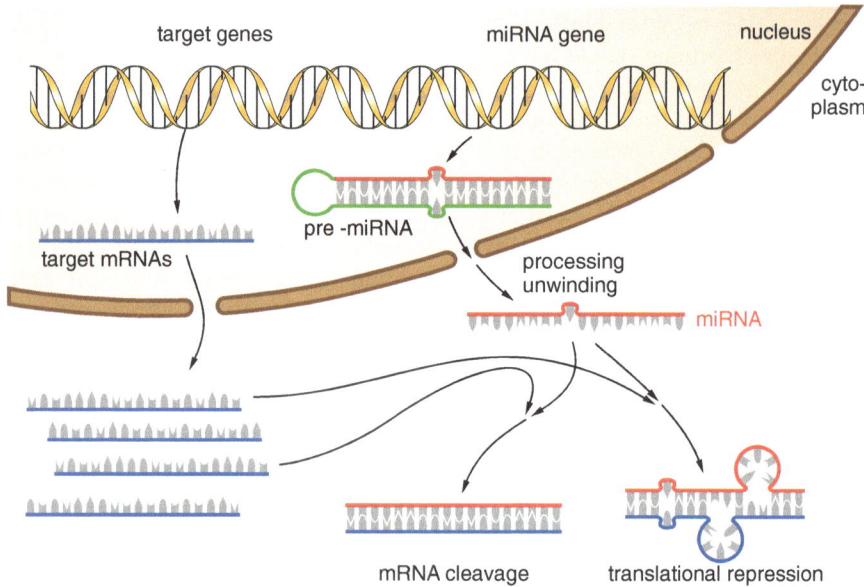

Fig. 3.23 Downregulation of gene expression by miRNA. MicroRNAs (miRNAs) are transcribed as large precursors edited in the nucleus to pre-miRNAs. In the cytoplasm they undergo additional processing to mature miRNA. By partial complementarity they can target a whole set of messenger RNAs leading to translational repression via conformational impediments. Perfect base pairing results in mRNA degradation (modified after Hammond 2005)

role of miRNAs in aging and the development of age-related disorders is found. For instance, the protein p53 that has been presented earlier as tumor suppressor protein and key checkpoint for the cell cycle, also regulates the expression of miRNAs. On the other hand p53 itself is an indirect target of regulation by miRNAs. Knowing that the expression of p53 is induced by insults affecting the integrity of the genomic DNA (e.g. UV and ionizing radiation) a miRNA-mediated down-regulation of p53 activity may have significant consequence on the development of cancer (He et al. 2007; Park et al. 2009; Zuckerman et al. 2009; Freeman and Espinosa 2013). More general, an upregulation of miRNAs with age and a global downregulation of miR-NAs in cancer has been described (Gonzalo 2010). Taken together, miRNAs belong to the toolset of epigenetics since they affect gene function independently of any changes in the nuclear genome sequence.

Lifestyle and environment affects aging: Different entities of human disorders are characterized by differences not only in the *genome, transcriptome* or *proteome* but also in the *epigenome*. Such pathologies that are affected by lifestyle and/or environmental factors which are manifested as epigenetic marks do not only include age-associated disease, but also, for instance, psychiatric conditions. Post traumatic stress disorders are of particular interest in this context but also the mother-foetus *in utero* relationship and its consequences for the development of disease in later life

(Schmidt et al. 2011; Galjaard et al. 2013). As the summary definition of epigenetics tells us "an epigenetic trait is a stable heritable phenotype resulting from changes in a chromosome without alterations in the DNA sequence" (Berger et al. 2009) it shows that exogenous factors and conditions can change DNA and its information content. Therefore, epigenetics is regarded to be the actual link between nature and nurture. Recent articles precipitate the consequences of this chemical sensitivity of the DNA and, inter alia, state "epigenetic patterns may change throughout one's life span, by an early life experience, environmental exposure or nutritional status. Epigenetic signatures influenced by the environment may determine our appearance, behavior, stress response, disease susceptibility, and even longevity" (Tammen et al. 2013). Environmental and behavioral factors that target the individual DNA can be summarized as personal lifestyle. And this, in fact, includes many factors on a daily basis such as the individual nutrition, experience of stress, sport and physical activity, the personal working style (e.g. shift working), and the consumption of drugs, alcohol and smoking. Many experimental studies now clearly show that such individual (soft) factors can directly influence the key epigenetic markers including DNA methylation, histone acetylation and the expression of miRNA. It is rather clear that the broad panel of individual lifestyle factors via epigenetic processes may affect human health (Alegría-Torres et al. 2011). In humans especially studies on identical twins highlight the importance of the individual life history for health, quality of life and longevity. Identical twins with identical DNA (sequence) may in fact differ quite massively at the level of epigenetic marks (Steves et al. 2012). Having the interplay between genes and environment in mind, in the context of our discussion here also aging needs to be considered. In fact, genetic changes (DNA sequence alterations, mutations, single nucleotide polymorphisms/SNPs) can be the reason for a predisposition for certain diseases and, therefore, cause a basic susceptibility (genetics). In addition, lifestyle and environmental factors may change DNA chemistry and disease disposition as well (epigenetics). Genes (genetics), environment (epigenetics) and their active interplay directly affect disease predisposition, the aging process and life span (Fig. 3.24).

3.9 A Holistic View: The Molecular Aging Matrix

To date, there is not one unifying theory of aging but many views that stand for their own. On the other hand most of the genes, molecules and pathways emphasized in those theories are linked and influence each other to some extent. In fact, initially aging has been understood as a stepwise and complex process of accumulation of damage in the key biomolecules of the cell (DNA, lipids and proteins). Aging and age-associated functional decline was simply explained as the result of a wear-and-tear process. But after decades of intensive molecular research cellular and organism aging is now acknowledged to be influenced and modulated by genes frequently called age-genes (a term that is not appropriate since these genes and their related proteins fulfill essential functions also during development and early in life)

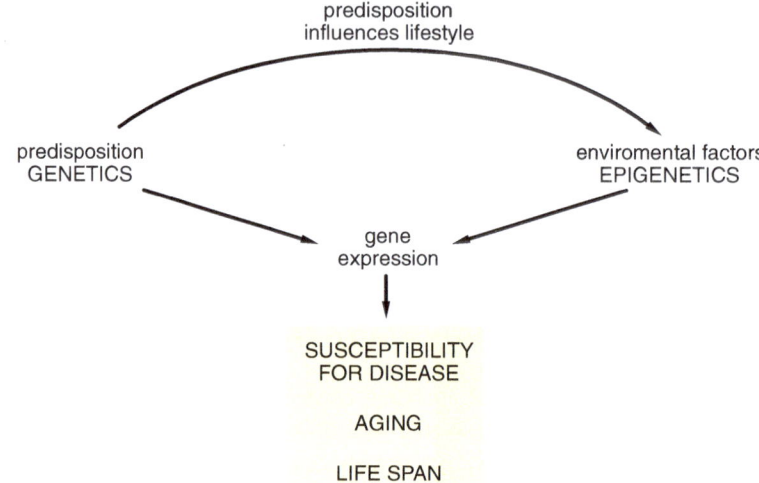

Fig. 3.24 Interplay of genetics and epigenetics. Genetics as the individual and fixed set of genes one carries is the key predisposition factor for disease susceptibility, life span and aging. Epigenetics as biochemical substrate of the environmental conditions affects these as well (modified after Holsboer 2007)

and age mechanisms that are evolutionary maintained. Their conservation suggests an important role for cells and organisms. Central examples of such mechanisms present over species boundaries are (1) insulin/insulin-like growth factor-driven signals, (2) mammalian target of Rapamycin (mTOR) signaling pathways, (3) oxidative stress or (4) mitochondrial activity. An individual manipulation of these genes and mechanisms leads to altered life span, at least in model organisms (Niccoli and Partridge 2012). On one hand, every single one of these age-related genes, proteins, and mechanisms is the core of a specific theory of aging. On the other hand, there is evidence showing the different individual pathways are interconnected. Examples are the nutrient-sensing insulin/IGF-1 system and the mTOR key regulator of autophagy, another example is that the extent of mitochondrial energy production is directly linked to the generation of mitochondria-derived oxidative stress and that there are many redox-regulated enzymes in the cell and oxidative stress-responsive genes, some of them belonging to the control of nutrient metabolism. A recent prominent example of such an interconnection is the analysis of epigenetic control mechanisms that regulate insulin-induced oxidative stress generation under high glucose condition linking these metabolic pathways also to epigenetics (Gupta and Tikoo 2012). In another very recent study focusing on mitochondria, the reactive oxygen species H_2O_2 that was introduced earlier as source of HO$^\cdot$ was found to function as a second messenger that is directly controlling the redox regulation of cell signaling and for its part is regulated by several cytosolic signaling pathways, such as the insulin/IGF1 function. Another intracellular signalling mechanism, the c-Jun N-terminal kinase (JNK) pathway, again is induced by oxidative stress and JNK itself

MOLECULAR AGING MATRIX

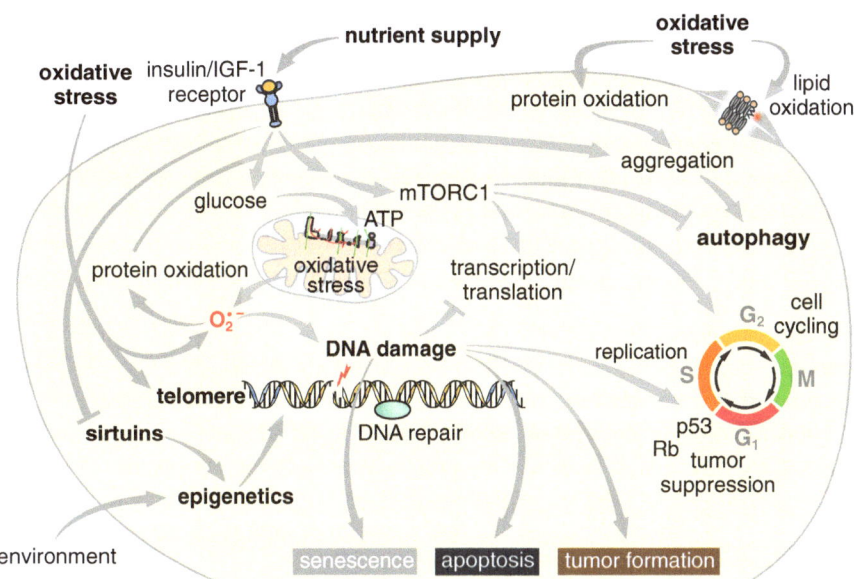

Fig. 3.25 An integrative theory of aging: the molecular aging matrix. Cellular aging is driven not by single genes or mechanisms but by an intimate network of influential factors: nutrient supply, oxidative stress, telomere length, epigenetics and sirtuin biochemistry, DNA damage and autophagy. Modulating one of these factors affects some if not all of the others, therefore, changing the whole cell biochemistry and physiology. We propose that the aforementioned central players are the determinants of the individual cell age status that set the stage for the transition of the cell into the aging process; we call this network the *molecular aging matrix*

via its translocation to the mitochondria can impair the generation of ATP. In turn the activity of insulin/IGF-1 and of JNK is connected, so that a metabolic triad is proposed consisting of (1) mitochondrial energy (and oxidative stress) generation, (2) the insulin/IGF-1 and (3) the JNK pathway, which is directly implicated in the aging of the brain and in the development of neurodegenerative disorders (Yin et al. 2013). Autophagy can be activated by DNA damage and the cells response to it and the role of autophagy in cell death following, for instance, genotoxic stress is well described (Surova and Zhivotovsky 2013). But the degradation of protein aggregates, organelle waste, defective mitochondria via autophagy can also contribute to genome stability assigning autophagy a novel role also in tumor suppression. As recently reviewed the stabilizing effect of autophagy on the genome can also be achieved by the elimination of damaged and defective nuclear parts (Vessoni et al. 2013) making autophagy in addition to mitochondria, the nucleus and membrane bound sensing receptors (e.g. insulin/IGF1-mediated mechanisms) a key signaling hub linking various age-related pathways. The key proteins, signaling mechanisms and pathways that have been shown to be interconnected, therefore, can be regarded to build a *molecular aging matrix* (Fig. 3.25).

In fact, evidence published demonstrating the interplay of the different key genes, proteins and pathways of aging is accumulating. Even more, obviously, there seems almost unlimited interdependence. Of course, other age researchers have connected various age-related pathways before but mostly focusing on a single pathway as the core such as, for instance, free radicals and oxidative stress. Shino Nemoto and Toren Finkel combined the activity of ROS with the function of the oncogene ras and sirtuins and some other players, herewith introducing a new version of the free radical theory of aging (Nemoto and Finkel 2004).

While finishing up this small book a review was published in the journal *Cell* summarizing what the authors called "the key hallmarks of aging". In fact in this elegant work exactly nine tentative hallmarks standing for what the authors call "common denominators of aging" (focusing on mammalian aging) are presented. Again, genomic instability, telomere attrition, epigenetic alterations, loss of proteostasis, deregulated nutrient sensing, mitochondrial dysfunction, cellular senescence, are labelled as hallmarks but also stem cell exhaustion and an altered intercellular communication (López-Otín et al. 2013). And as pointed out in our present discussion also in this article the effort to dissect the interconnectedness between these formulated hallmarks and their relative and individual contributions to aging is addressed as the actual key future challenge of aging research.

"Aging is unquestioningly complex" (Kirkwood 2011): What has not been discussed here so far is the link between evolution and aging. "Aging is usually defined as the progressive loss of function accompanied by decreasing fertility and increasing mortality with advancing age" (Kirkwood and Austad 2000). Considering different species *evolutionary theories of aging* (there is more than just one theory since different theories focus on different evolutionary aspects) have evolved that explain why aging occurs in principal, by focusing on the fertility of the individual species and evolutionary factors. As summarized in a landmark review on this topic by Kirkwood and Austad, evolutionary theories of aging can predict (sentences in italics taken from the original paper by Kirkwood and Austad 2000): 1. *Specific genes selected to promote aging are unlikely to exist.* This means that there is no genetic module of aging that carries the information on the life span of an individual species. Consequently, there is not *the* aging gene that can be modulated. 2. *Aging is not programmed but results largely from accumulation of somatic damage, owing to limited investments in maintenance and repair. Longevity is thus regulated by genes controlling levels of activities such as DNA repair and antioxidant defence.* There exists a large panel of genes that code for defence and repair proteins and enzymes that as a whole affect the aging process and, in particular, resistance to age disorders, especially cancer. 3. *In addition, there may be adverse gene actions at older ages arising either from purely deleterious genes that escape the forces of natural selection or from pleiotropic genes that trade benefit at an early age against harm at older ages.* One has to keep in mind that selection takes place at the level of reproduction and that there is no selective pressure against aging as it occurs after the reproductive phase of an individual. Considering evolutionary theories of aging it can be simply said that when the pressure of evolution is high, organisms die before they can age and so, ultimately, aging results from the decline in natural selection.

So, considering the plethora of data linking certain genes, pathways and conditions with aging after all what we are mainly left with and what can everyone perhaps agree on is the significant impact of the interplay of *nature and nurture* and the fact that the aging process and the aging phenotype are the result of a complex interaction between genetic and environmental factors. In a similar manner this has been formulated already about three decades ago in saying "Finally, we hypothesize that the factors which determine maximum longevity of individuals in a population are the rate of occurrence of DNA damage, the rate of DNA repair, the degree of cellular redundancy, and the extent of exposure to stress" (Gensler and Bernstein 1981). And even more complex is the mutual influence of aging on age-related disease and vice versa, which will be shortly addressed in the final chapter.

References

Adwan L, Zawia NH (2013) Epigenetics: a novel therapeutic approach for the treatment of Alzheimer's disease. Pharmacol Ther 139(1):41–50

Alberts B, Johnson A, Walter P, Lewis J, Raff M, Roberts K (2007) Molecular biology of the cell, 5th revised edn. Taylor & Francis, New York

Alegría-Torres JA, Baccarelli A, Bollati V (2011) Epigenetics and lifestyle. Epigenomics 3(3):267–277

Alexander P (1967) The role of DNA lesions in processes leading to aging in mice. Symp Soc Exp Biol 21:29–50

Amm I, Sommer T, Wolf DH (2013) Protein quality control and elimination of protein waste: the role of the ubiquitin-proteasome system. Biochim Biophys Acta [Epub ahead of print]

Anselmi B, Conconi M, Veyrat-Durebex C, Turlin E, Biville F, Alliot J, Friguet B (1998) Dietary self-selection can compensate an age-related decrease of rat liver 20 S proteasome activity observed with standard diet. J Gerontol A Biol Sci Med Sci 53(3):B173–B179

Atzmon G, Cho M, Cawthon RM, Budagov T, Katz M, Yang X, Siegel G, Bergman A, Huffman DM, Schechter CB, Wright WE, Shay JW, Barzilai N, Govindaraju DR, Suh Y (2010) Evolution in health and medicine Sackler colloquium: genetic variation in human telomerase is associated with telomere length in Ashkenazi centenarians. Proc Natl Acad Sci U S A 107(Suppl 1):1710–1717

Austad SN (2010) Methusaleh's Zoo: how nature provides us with clues for extending human health span. J Comp Pathol 142(Suppl 1):S10–S21

Bae YS, Oh H, Rhee SG, Yoo YD (2011) Regulation of reactive oxygen species generation in cell signaling. Mol Cells 32(6):491–509

Balaban RS, Nemoto S, Finkel T (2005) Mitochondria, oxidants, and aging. Cell 120(4):483–495

Barneda-Zahonero B, Parra M (2012) Histone deacetylases and cancer. Mol Oncol 6(6):579–589

Bartke A (2011) Single-gene mutations and healthy ageing in mammals. Philos Trans R Soc Lond B Biol Sci 366(1561):28–34

Bártová E, Krejčí J, Harnicarová A, Galiová G, Kozubek S (2008) Histone modifications and nuclear architecture: a review. J Histochem Cytochem 56(8):711–721

Beauharnois JM, Bolívar BE, Welch JT (2013) Sirtuin 6: a review of biological effects and potential therapeutic properties. Mol Biosyst 9(7):1789–1806

Behl C, Davis JB, Lesley R, Schubert D (1994) Hydrogen peroxide mediates amyloid beta protein toxicity. Cell 77(6):817–827

Behl C, Moosmann B (2002) Oxidative nerve cell death in Alzheimer's disease and stroke: antioxidants as neuroprotective compounds. Biol Chem 383(3–4):521–536

Behl C (2012) Brain aging and late-onset Alzheimer's disease: many open questions. Int Psychogeri-
atr 24(Suppl 1):S3–S9
Behl C, Moosmann B (2008) Molekulare Mechanismen des Alterns. Über das Altern der Zellen
und den Einfluss von oxidativem Stress auf den Alternsprozess. In: Staudinger UM, Häfner H
(eds) Was ist Alter(n)? Neue Antworten auf eine scheinbar einfache Frage, pp 9–32. Spinger,
Berlin [Schriften der Mathematisch-naturwissenschaftlichen Klasse der Heidelberger Akademie
der Wissenschaften, Nr. 18, 2008]
Ben-Avraham D, Muzumdar RH, Atzmon G (2012) Epigenetic genome-wide association methyla-
tion in aging and longevity. Epigenomics 4(5):503–509
Bender A, Hajieva P, Moosmann B (2008) Adaptive antioxidant methionine accumulation in res-
piratory chain complexes explains the use of a deviant genetic code in mitochondria. Proc Natl
Acad Sci U S A 105(43):16496–16501
Berger SL, Kouzarides T, Shiekhattar R, Shilatifard A (2009) An operational definition of epige-
netics. Genes Dev 23(7):781–783
Bjornsson HT, Sigurdsson MI, Fallin MD, Irizarry RA, Aspelund T, Cui H, Yu W, Rongione MA,
Ekström TJ, Harris TB, Launer LJ, Eiriksdottir G, Leppert MF, Sapienza C, Gudnason V, Feinberg
AP (2008) Intra-individual change over time in DNA methylation with familial clustering. JAMA
299(24):2877–2883
Blüher M, Kahn BB, Kahn CR (2003) Extended longevity in mice lacking the insulin receptor in
adipose tissue. Science 299(5606):572–574
Bourzac K (2012) Interventions: live long and prosper. Nature 492(7427):S18–S20
Branzei D, Foiani M (2008) Regulation of DNA repair throughout the cell cycle. Nat Rev Mol Cell
Biol 9(4):297–308
Brown MK, Naidoo N (2012) The endoplasmic reticulum stress response in aging and age-related
diseases. Front Physiol 3:263
Brown-Borg HM, Borg KE, Meliska CJ, Bartke A (1996) Dwarf mice and the ageing process.
Nature 384(6604):33
Brown-Borg HM, Bartke A (2012) GH and IGF1: roles in energy metabolism of long-living GH
mutant mice. J Gerontol A Biol Sci Med Sci 67(6):652–660
Bukau B, Weissman J, Horwich A (2006) Molecular chaperones and protein quality control. Cell
125(3):443–451
Burgess RJ, Zhang Z (2010) Histones, histone chaperones and nucleosome assembly. Protein Cell
1(7):607–612
Burtner CR, Kennedy BK (2010) Progeria syndromes and ageing: what is the connection? Nat Rev
Mol Cell Biol 11(8):567–578
Busque L, Mio R, Mattioli J, Brais E, Blais N, Lalonde Y, Maragh M, Gilliland DG (1996) Nonran-
dom X-inactivation patterns in normal females: lyonization ratios vary with age. Blood 88(1):59–
65
Carafa V, Nebbioso A, Altucci L (2012) Sirtuins and disease: the road ahead. Front Pharmacol 3:4
Casorelli I, Bossa C, Bignami M (2012) DNA damage and repair in human cancer: molecular
mechanisms and contribution to therapy-related leukemias. Int J Environ Res Public Health
9(8):2636–2657
Cech TR (2004) Beginning to understand the end of the chromosome. Cell 116(2):273–279
Chavez E, Vulto I, Lansdorp PM (2009) Telomere length in Hutchinson-Gilford progeria syndrome.
Mech Ageing Dev 130(6):377–383
Chen Y, Klionsky DJ (2011) The regulation of autophagy—unanswered questions. J Cell Sci 124(Pt
2):161–170
Chevanne M, Calia C, Zampieri M, Cecchinelli B, Caldini R, Monti D, Bucci L, Franceschi C,
Caiafa P (2007) Oxidative DNA damage repair and parp 1 and parp 2 expression in Epstein-
Barr virus-immortalized B lymphocyte cells from young subjects, old subjects, and centenarians.
Rejuvenation Res 10(2):191–204

Chouliaras L, van den Hove DL, Kenis G, Keitel S, Hof PR, van Os J, Steinbusch HW, Schmitz C, Rutten BP (2012) Prevention of age-related changes in hippocampal levels of 5-methylcytidine by caloric restriction. Neurobiol Aging 33(8):1672–1681

Clancy DJ, Gems D, Harshman LG, Oldham S, Stocker H, Hafen E, Leevers SJ, Partridge L (2001) Extension of life-span by loss of CHICO, a Drosophila insulin receptor substrate protein. Science 292(5514):104–106

Cleaver JE, Lam ET, Revet I (2009) Disorders of nucleotide excision repair: the genetic and molecular basis of heterogeneity. Nat Rev Genet 10(11):756–768

Clement AB, Gamerdinger M, Tamboli IY, Lütjohann D, Walter J, Greeve I, Gimpl G, Behl C (2009) Adaptation of neuronal cells to chronic oxidative stress is associated with altered cholesterol and sphingolipid homeostasis and lysosomal function. J Neurochem 111(3):669–682

Clement AB, Gimpl G, Behl C (2010) Oxidative stress resistance in hippocampal cells is associated with altered membrane fluidity and enhanced nonamyloidogenic cleavage of endogenous amyloid precursor protein. Free Radic Biol Med 48(9):1236–1241

Cline SD (2012) Mitochondrial DNA damage and its consequences for mitochondrial gene expression. Biochim Biophys Acta 1819(9–10):979–991

Colman RJ, Anderson RM, Johnson SC, Kastman EK, Kosmatka KJ, Beasley TM, Allison DB, Cruzen C, Simmons HA, Kemnitz JW, Weindruch R (2009) Caloric restriction delays disease onset and mortality in rhesus monkeys. Science 325(5937):201–204

Corey DR (2009) Telomeres and telomerase: from discovery to clinical trials. Chem Biol 16(12):1219–1223

Cornaro L (2005) English translation by Butler WF (1903) The art of living long. Springer, New York

Couzin-Frankel J (2011) Genetics. Aging genes: the sirtuin story unravels. Science 334(6060):1194–1198

Cuervo AM, Dice JF (2000) Age-related decline in chaperone-mediated autophagy. J Biol Chem 275(40):31505–31513

Culotta E, Koshland DE Jr (1992) NO news is good news. Science 258(5090):1862–1865

Curtin NJ (2012) DNA repair dysregulation from cancer driver to therapeutic target. Nat Rev Cancer 12(12):801–817

D'Aquila P, Rose G, Bellizzi D, Passarino G (2013) Epigenetics and aging. Maturitas 74(2):130–136

Dasuri K, Zhang L, Keller JN (2013) Oxidative stress, neurodegeneration, and the balance of protein degradation and protein synthesis. Free Radic Biol Med 62:170–185

David DC, Ollikainen N, Trinidad JC, Cary MP, Burlingame AL, Kenyon C (2010) Widespread protein aggregation as an inherent part of aging in *C. elegans*. PLoS Biol 8:e1000450

David DC (2012) Aging and the aggregating proteome. Front Genet 3:247

Decker ML, Chavez E, Vulto I, Lansdorp PM (2009) Telomere length in Hutchinson-Gilford progeria syndrome. Mech Ageing Dev 130(6):377–383

Dhurandhar EJ, Allison DB, van Groen T, Kadish I (2013) Hunger in the absence of caloric restriction improves cognition and attenuates Alzheimer's disease pathology in a mouse model. PLoS One 8(4):e60437

Dikic I, Johansen T, Kirkin V (2010) Selective autophagy in cancer development and therapy. Cancer Res 70(9):3431–3434

Dobashi Y, Watanabe Y, Miwa C, Suzuki S, Koyama S (2011) Mammalian target of rapamycin: a central node of complex signaling cascades. Int J Clin Exp Pathol 4(5):476–495

Dong S, Duan Y, Hu Y, Zhao Z (2012) Advances in the pathogenesis of Alzheimer's disease: a re-evaluation of amyloid cascade hypothesis. Transl Neurodegener 1(1):18

Donmez G, Wang D, Cohen DE, Guarente L (2010) SIRT1 suppresses beta-amyloid production by activating the alpha-secretase gene ADAM10. Cell 142(2):320–332 (Erratum in: Cell 142(3):494–495)

Dorman JB, Albinder B, Shroyer T, Kenyon C (1995) The age-1 and daf-2 genes function in a common pathway to control the lifespan of *Caenorhabditis elegans*. Genetics 141(4):1399–1406

Dunlop RA, Brunk UT, Rodgers KJ (2009) Oxidized proteins: mechanisms of removal and consequences of accumulation. IUBMB Life 61(5):522–527

De Duve C, Wattiaux R (1966) Functions of lysosomes. Annu Rev Physiol 28:435–492

Ewbank JJ (2006) Signaling in the immune response (23 Jan 2006). In: WormBook (ed) The *C. elegans* research community, WormBook. doi:10.1895/wormbook.1.83.1, http://www.wormbook.
org

Fontana L, Partridge L, Longo VD (2010) Extending healthy life span-from yeast to humans.
Science 328(5976):321–326

Foster DA, Yellen P, Xu L, Saqcena M (2010) Regulation of G1 cell cycle progression: distinguishing
the restriction point from a nutrient-sensing cell growth checkpoint(s). Genes Cancer 1(11):1124–
1131

Fraga MF, Esteller M (2007) Epigenetics and aging: the targets and the marks. Trends Genet
23(8):413–418

Fredrickson EK, Gardner RG (2012) Selective destruction of abnormal proteins by ubiquitin-
mediated protein quality control degradation. Semin Cell Dev Biol 23(5):530–537

Freeman JA, Espinosa JM (2013) The impact of post-transcriptional regulation in the p53 network.
Brief Funct Genomics 12(1):46–57

Freitas AA, de Magalhães JP (2011) A review and appraisal of the DNA damage theory of ageing.
Mutat Res 728(1–2):12–22

Friedman DB, Johnson TE (1988) A mutation in the age-1 gene in *Caenorhabditis elegans* lengthens
life and reduces hermaphrodite fertility. Genetics 118(1):75–86

Galjaard S, Devlieger R, Van Assche FA (2013) Fetal growth and developmental programming. J
Perinat Med 41(1):101–105

Gamerdinger M, Hajieva P, Kaya AM, Wolfrum U, Hartl FU, Behl C (2009) Protein quality control
during aging involves recruitment of the macroautophagy pathway by BAG3. EMBO J 28(7):889–
901

Gamerdinger M, Carra S, Behl C (2011b) Emerging roles of molecular chaperones and co-
chaperones in selective autophagy: focus on BAG proteins. J Mol Med (Berl) 89(12):1175–1182

Gamerdinger M, Kaya AM, Wolfrum U, Clement AM, Behl C (2011a) BAG3 mediates chaperone-
based aggresome-targeting and selective autophagy of misfolded proteins. EMBO Rep 12(2):149–
56

Gensler HL, Bernstein H (1981) DNA damage as the primary cause of aging. Q Rev Biol 56(3):279–
303

Germann MW, Johnson CN, Spring AM (2012) Recognition of damaged DNA: structure and
dynamic markers. Med Res Rev 32(3):659–683

Gkogkolou P, Böhm M (2012) Advanced glycation end products: key players in skin aging? Der-
matoendocrinol 4(3):259–270

González-Suárez E, Geserick C, Flores JM, Blasco MA (2005) Antagonistic effects of telomerase
on cancer and aging in K5-mTert transgenic mice. Oncogene 24(13):2256–2270

Gonzalo S (2010) Epigenetic alterations in aging. J Appl Physiol 109(2):586–597

Gredilla R, Garm C, Stevnsner T (2012) Nuclear and mitochondrial DNA repair in selected eukaryotic aging model systems. Oxid Med Cell Longev 2012:282438

Greer EL, Maures TJ, Hauswirth AG, Green EM, Leeman DS, Maro GS, Han S, Banko MR,
Gozani O, Brunet A (2010) Members of the H3K4 trimethylation complex regulate lifespan in a
germline-dependent manner in *C. elegans*. Nature 466(7304):383–387

Greeve I, Hermans-Borgmeyer I, Brellinger C, Kasper D, Gomez-Isla T, Behl C, Levkau B,
Nitsch RM (2000) The human DIMINUTO/DWARF1 homolog seladin-1 confers resistance to
Alzheimer's disease-associated neurodegeneration and oxidative stress. J Neurosci 20(19):7345–
7352

Greider CW, Blackburn EH (1985) Identification of a specific telomere terminal transferase activity
in Tetrahymena extracts. Cell 43(2 Pt 1):405–413

Grillari J, Grillari-Voglauer R (2010) Novel modulators of senescence, aging, and longevity: small
non-coding RNAs enter the stage. Exp Gerontol 45(4):302–311

Guarente L (2011) Franklin H. Epstein lecture: sirtuins, aging, and medicine. N Engl J Med 364(23):2235–2244

Guarente L (2013) Calorie restriction and sirtuins revisited. Genes Dev 27(19):2072–2085

Gupta J, Tikoo K (2012) Involvement of insulin-induced reversible chromatin remodeling in altering the expression of oxidative stress-responsive genes under hyperglycemia in 3T3-L1 preadipocytes. Gene 504(2):181–191

Halliwell B, Gutteridge JMC (1999) Free radicals in biology and medicine, 3rd edn. Clarendon Press, Oxford

Hammond SM (2005) Dicing and slicing: the core machinery of the RNA interference pathway. FEBS Lett 579(26):5822–5829

Harley CB, Sherwood SW (1997) Telomerase, checkpoints and cancer. Cancer Surv 29:263–284

Harman D (1956) Aging: a theory based on free radical and radiation chemistry. J Gerontol 11(3):298–300

Harman D (1972) The biologic clock: the mitochondria? J Am Geriatr Soc 20(4):145–147

Harman D (2009) About "origin and evolution of the free radical theory of aging: a brief personal history, 1954–2009". Biogerontology 10(6):783

Harrison DE, Strong R, Sharp ZD, Nelson JF, Astle CM, Flurkey K, Nadon NL, Wilkinson JE, Frenkel K, Carter CS, Pahor M, Javors MA, Fernandez E, Miller RA (2009) Rapamycin fed late in life extends lifespan in genetically heterogeneous mice. Nature 460(7253):392–395

Hartl FU, Hayer-Hartl M (2002) Molecular chaperones in the cytosol: from nascent chain to folded protein. Science 295:1852–1858

He L, He X, Lowe SW, Hannon GJ (2007) MicroRNAs join the p53 network—another piece in the tumour-suppression puzzle. Nat Rev Cancer 7(11):819–822

He XJ, Chen T, Zhu JK (2011) Regulation and function of DNA methylation in plants and animals. Cell Res 21(3):442–465

Hecht SS (2012) Lung carcinogenesis by tobacco smoke. Int J Cancer 131(12):2724–2732

Heilbronn LK, de Jonge L, Frisard MI, DeLany JP, Larson-Meyer DE, Rood J, Nguyen T, Martin CK, Volaufova J, Most MM, Greenway FL, Smith SR, Deutsch WA, Williamson DA, Ravussin E, Pennington CALERIE Team (2006) Effect of 6-month calorie restriction on biomarkers of longevity, metabolic adaptation, and oxidative stress in overweight individuals: a randomized controlled trial. JAMA 295(13):1539–48 (Erratum in: JAMA 295(21):2482)

Heydari AR, You S, Takahashi R, Gutsmann-Conrad A, Sarge KD, Richardson A (2000) Age-related alterations in the activation of heat shock transcription factor 1 in rat hepatocytes. Exp Cell Res 256:83–93

Hochfeld WE, Lee S, Rubinsztein DC (2013) Therapeutic induction of autophagy to modulate neurodegenerative disease progression. Acta Pharmacol Sin 34(5):600–604

Hoeijmakers JH (2001) Genome maintenance mechanisms for preventing cancer. Nature 411(6835):366–374

Holsboer F (2007) Altersbedingte Erkrankungen: Das Wechselspiel von Veranlagung und Lebensweise. In: Gruss P (ed) Die Zukunft des Alterns, pp 163–191. C.H. Beck, München

Holzenberger M, Dupont J, Ducos B, Leneuve P, Géloën A, Even PC, Cervera P, Le Bouc Y (2003) IGF-1 receptor regulates lifespan and resistance to oxidative stress in mice. Nature 421(6919):182–187

Horcajada MN, Offord E (2012) Naturally plant-derived compounds: role in bone anabolism. Curr Mol Pharmacol 5(2):205–218

Howitz KT, Bitterman KJ, Cohen HY, Lamming DW, Lavu S, Wood JG, Zipkin RE, Chung P, Kisielewski A, Zhang LL, Scherer B, Sinclair DA (2003) Small molecule activators of sirtuins extend *Saccharomyces cerevisiae* lifespan. Nature 425(6954):191–196

Hsu AL, Murphy CT, Kenyon C (2003) Regulation of aging and age-related disease by DAF-16 and heat-shock factor. Science 300:1142–1145

Humphreys V, Martin RM, Ratcliffe B, Duthie S, Wood S, Gunnell D, Collins AR (2007) Age-related increases in DNA repair and antioxidant protection: a comparison of the Boyd Orr Cohort of elderly subjects with a younger population sample. Age Ageing 36(5):521–526

Ibáñez-Ventoso C, Driscoll M (2009) MicroRNAs in *C. elegans* aging: molecular insurance for robustness? Curr Genomics 10(3):144–153

Jeck WR, Siebold AP, Sharpless NE (2012) Review: a meta-analysis of GWAS and age-associated diseases. Aging Cell 11(5):727–731

Jena NR (2012) DNA damage by reactive species: mechanisms, mutation and repair. J Biosci 37(3):503–517

Jeppesen DK, Bohr VA, Stevnsner T (2011) DNA repair deficiency in neurodegeneration. Prog Neurobiol 94(2):166–200

Jones QR, Warford J, Rupasinghe HP, Robertson GS (2012) Target-based selection of flavonoids for neurodegenerative disorders. Trends Pharmacol Sci 33(11):602–610

Jung T, Bader N, Grune T (2007) Lipofuscin: formation, distribution, and metabolic consequences. Ann N Y Acad Sci 1119:97–111

Jung HJ, Suh Y (2012) MicroRNA in aging: from discovery to biology. Curr Genomics 13(7):548–557

Kaarniranta K, Salminen A, Eskelinen EL, Kopitz J (2009) Heat shock proteins as gatekeepers of proteolytic pathways—implications for age-related macular degeneration (AMD). Ageing Res Rev 8(2):128–139

Kaelin WG Jr, McKnight SL (2013) Influence of metabolism on epigenetics and disease. Cell 153(1):56–69

Kamileri I, Karakasilioti I, Garinis GA (2012) Nucleotide excision repair: new tricks with old bricks. Trends Genet 28(11):566–573

Kanungo J (2013) DNA-dependent protein kinase and DNA repair: relevance to Alzheimer's disease. Alzheimers Res Ther 5(2):13

Kenyon C, Chang J, Gensch E, Rudner A, Tabtiang R (1993) A *C. elegans* mutant that lives twice as long as wild type. Nature 366(6454):461–464

Kern A, Ackermann B, Clement AM, Duerk H, Behl C (2010) HSF1-controlled and age-associated chaperone capacity in neurons and muscle cells of *C. elegans*. PLoS One 5(1):e8568

Kim YJ, Wilson DM 3rd (2012) Overview of base excision repair biochemistry. Curr Mol Pharmacol 5(1):3–13

Kim HS, Patel K, Muldoon-Jacobs K, Bisht KS, Aykin-Burns N, Pennington JD, van der Meer R, Nguyen P, Savage J, Owens KM, Vassilopoulos A, Ozden O, Park SH, Singh KK, Abdulkadir SA, Spitz DR, Deng CX, Gius D (2010) SIRT3 is a mitochondria-localized tumor suppressor required for maintenance of mitochondrial integrity and metabolism during stress. Cancer Cell 17(1):41–52

Kimura KD, Tissenbaum HA, Liu Y, Ruvkun G (1997) daf-2, an insulin receptor-like gene that regulates longevity and diapause in *Caenorhabditis elegans*. Science 277(5328):942–946.

Kirkwood TB (2011) Systems biology of ageing and longevity. Philos Trans R Soc Lond B Biol Sci 366(1561):64–70

Kirkwood TB, Austad SN (2000) Why do we age? Nature 408(6809):233–238

Klapper W, Parwaresch R, Krupp G (2001) Telomere biology in human aging and aging syndromes. Mech Ageing Dev 122(7):695–712

Koshland DE Jr (1992) The molecule of the year. Science 258(5090):1861

Krokan HE, Bjørås M (2013) Base excision repair. Cold Spring Harb Perspect Biol 5(4):a012583

Kuro-o M (2012) Klotho in health and disease. Curr Opin Nephrol Hypertens 21(4):362–368

Lamy E, Goetz V, Erlacher M, Herz C, Mersch-Sundermann V (2013) hTERT: another brick in the wall of cancer cells. Mutat Res 752(2):119–128

Lee RC, Feinbaum RL, Ambros V (1993) The *C. elegans* heterochronic gene lin-4 encodes small RNAs with antisense complementarity to lin-14. Cell 75(5):843–854

van Leeuwen FW, de Kleijn DP, van den Hurk HH, Neubauer A, Sonnemans MA, Sluijs JA, Köycü S, Ramdjielal RD, Salehi A, Martens GJ, Grosveld FG, Peter J, Burbach H, Hol EM (1998) Frameshift mutants of beta amyloid precursor protein and ubiquitin-B in Alzheimer's and down patients. Science 279(5348):242–247

Lehmann AR, McGibbon D, Stefanini M (2011) Xeroderma pigmentosum. Orphanet J Rare Dis 6:70

Li N, Karin M (1999) Is NF-kappaB the sensor of oxidative stress? FASEB J 13(10):1137–1143

Lieber MR, Ma Y, Pannicke U, Schwarz K (2003) Mechanism and regulation of human non-homologous DNA end-joining. Nat Rev Mol Cell Biol 4(9):712–720

Lieber MR (2010) The mechanism of double-strand DNA break repair by the nonhomologous DNA end-joining pathway. Annu Rev Biochem 79:181–211

Liochev SI (2013) Reactive oxygen species and the free radical theory of aging. Free Radic Biol Med 60:1–4

Liscic RM, Breljak D (2011) Molecular basis of amyotrophic lateral sclerosis. Prog Neuropsychopharmacol Biol Psychiatry 35(2):370–372

Lombard DB, Chua KF, Mostoslavsky R, Franco S, Gostissa M, Alt FW (2005) DNA repair, genome stability, and aging. Cell 120(4):497–512

López-Otín C, Blasco MA, Partridge L, Serrano M, Kroemer G (2013) The hallmarks of aging. Cell 153(6):1194–1217

Lu T, Pan Y, Kao SY, Li C, Kohane I, Chan J, Yankner BA (2004) Gene regulation and DNA damage in the ageing human brain. Nature 429(6994):883–891

Marmorstein R, Roth SY (2001) Histone acetyltransferases: function, structure, and catalysis. Curr Opin Genet Dev 11(2):155–161

Marquardt JU, Fischer K, Baus K, Kashyap A, Ma S, Krupp M, Linke M, Teufel A, Zechner U, Strand D, Thorgeirsson SS, Galle PR, Strand S (2013) SIRT6 dependent genetic and epigenetic alterations are associated with poor clinical outcome in HCC patients. Hepatology 58(3):1054–1064

Masters CL, Selkoe DJ (2012) Biochemistry of amyloid β-protein and amyloid deposits in Alzheimer disease. Cold Spring Harb Perspect Med 2(6):a006262

Masui R, Kuramitsu S (2010) Molecular mechanisms of the whole DNA repair system: a comparison of bacterial and eukaryotic systems. J Nucleic Acids 2010:179594

Mattison JA, Roth GS, Beasley TM, Tilmont EM, Handy AM, Herbert RL, Longo DL, Allison DB, Young JE, Bryant M, Barnard D, Ward WF, Qi W, Ingram DK, de Cabo R (2012) Impact of caloric restriction on health and survival in rhesus monkeys from the NIA study. Nature 489(7415):318–321

Mattson MP (2009) Roles of the lipid peroxidation product 4-hydroxynonenal in obesity, the metabolic syndrome, and associated vascular and neurodegenerative disorders. Exp Gerontol 44(10):625–633

Mayer MP, Bukau B (2005) Hsp70 chaperones: cellular functions and molecular mechanism. Cell Mol Life Sci 62(6):670–684

Ma D, Zhu W, Hu S, Yu X, Yang Y (2013) Association between oxidative stress and telomere length in type 1 and type 2 diabetic patients. J Endocrinol Invest [Epub ahead of print]

McCay CM (2000) Is longevity compatible with optimum growth? Science 77(2000):410–411

McCollum AK, Casagrande G, Kohn EC (2010) Caught in the middle: the role of Bag3 in disease. Biochem J 425:e1–e3

McCord JM, Fridovich I (2013) Superoxide dismutases: you've come a long way, baby. Antioxid Redox Signal [Epub ahead of print]

McCord RA, Michishita E, Hong T, Berber E, Boxer LD, Kusumoto R, Guan S, Shi X, Gozani O, Burlingame AL, Bohr VA, Chua KF (2009) SIRT6 stabilizes DNA-dependent protein kinase at chromatin for DNA double-strand break repair. Aging 1(1):109–121

McCord JM, Fridovich I (1969) Superoxide dismutase. An enzymic function for erythrocuprein (hemocuprein). J Biol Chem 244(22):6049–6055

McGuinness D, McGuinness DH, McCaul JA, Shiels PG (2011) Sirtuins, bioageing, and cancer. J Aging Res 2011:235754

McKinnon PJ (2012) ATM and the molecular pathogenesis of ataxia telangiectasia. Annu Rev Pathol 7:303–321

Meng F, Yao D, Shi Y, Kabakoff J, Wu W, Reicher J, Ma Y, Moosmann B, Masliah E, Lipton SA, Gu Z (2011) Oxidation of the cysteine-rich regions of parkin perturbs its E3 ligase activity and contributes to protein aggregation. Mol Neurodegener 6:34

Merksamer PI, Liu Y, He W, Hirschey MD, Chen D, Verdin E (2013) The sirtuins, oxidative stress and aging: an emerging link. Aging 5(3):144–150

Michael R, Bron AJ (2011) The ageing lens and cataract: a model of normal and pathological ageing. Philos Trans R Soc Lond B Biol Sci 366(1568):1278–1292

Mocko JB, Kern A, Moosmann B, Behl C, Hajieva P (2010) Phenothiazines interfere with dopaminergic neurodegeneration in *Caenorhabditis elegans* models of Parkinson's disease. Neurobiol Dis 40(1):120–129

Mogk A, Schmidt R, Bukau B (2007) The N-end rule pathway for regulated proteolysis: prokaryotic and eukaryotic strategies. Trends Cell Biol 17(4):165–172

Moore JK, Haber JE (1996) Cell cycle and genetic requirements of two pathways of nonhomologous end-joining repair of double-strand breaks in *Saccharomyces cerevisiae*. Mol Cell Biol 16(5):2164–2173

Moosmann B, Behl C (1999) The antioxidant neuroprotective effects of estrogens and phenolic compounds are independent from their estrogenic properties. Proc Natl Acad Sci U S A 96(16):8867–8872

Moosmann B, Behl C (2002) Antioxidants as treatment for neurodegenerative disorders. Expert Opin Investig Drugs 11(10):1407–1435

Moosmann B, Behl C (2008) Mitochondrially encoded cysteine predicts animal lifespan. Aging Cell 7(1):32–46

Morawe T, Hiebel C, Kern A, Behl C (2012) Protein homeostasis aging and Alzheimer's disease. Mol Neurobiol 46(1):41–54

Morris JZ, Tissenbaum HA, Ruvkun G (1996) A phosphatidylinositol-3-OH kinase family member regulating longevity and diapause in *Caenorhabditis elegans*. Nature 382(6591):536–539

Mostoslavsky R, Chua KF, Lombard DB, Pang WW, Fischer MR, Gellon L, Liu P, Mostoslavsky G, Franco S, Murphy MM, Mills KD, Patel P, Hsu JT, Hong AL, Ford E, Cheng HL, Kennedy C, Nunez N, Bronson R, Frendewey D, Auerbach W, Valenzuela D, Karow M, Hottiger MO, Hursting S, Barrett JC, Guarente L, Mulligan R, Demple B, Yancopoulos GD, Alt FW (2006) Genomic instability and aging-like phenotype in the absence of mammalian SIRT6. Cell 124(2):315–329

Moulson CL, Fong LG, Gardner JM, Farber EA, Go G, Passariello A, Grange DK, Young SG, Miner JH (2007) Increased progerin expression associated with unusual LMNA mutations causes severe progeroid syndromes. Hum Mutat 28(9):882–889

Müller-Esterl W (2011) Biochemie: Eine Einführung für Mediziner und Naturwissenschaftler. Spektrum Akademischer Verlag, 2. Auflage

Murabito JM, Yuan R, Lunetta KL (2012) The search for longevity and healthy aging genes: insights from epidemiological studies and samples of long-lived individuals. J Gerontol A Biol Sci Med Sci 67(5):470–479

Nauseef WM (1999) The NADPH-dependent oxidase of phagocytes. Proc Assoc Am Physicians 111(5):373–382

Nemoto S, Finkel T (2004) Ageing and the mystery at Arles. Nature 429(6988):149–152

Niccoli T, Partridge L (2012) Ageing as a risk factor for disease. Curr Biol 22(17):R741–752

Niedernhofer LJ (2008) Tissue-specific accelerated aging in nucleotide excision repair deficiency. Mech Ageing Dev 129(7–8):408–415

De Oliveira RM, Sarkander J, Kazantsev AG, Outeiro TF (2012) SIRT2 as a therapeutic target for age-related disorders. Front Pharmacol 3:82

Olovnikov AM (1996) Telomeres, telomerase, and aging: origin of the theory. Exp Gerontol 31(4):443–448

Pamplona R, Barja G (2006) Mitochondrial oxidative stress, aging and caloric restriction: the protein and methionine connection. Biochim Biophys Acta 1757(5–6):496–508

Pan MH, Lai CS, Tsai ML, Wu JC, Ho CT (2012) Molecular mechanisms for anti-aging by natural dietary compounds. Mol Nutr Food Res 56(1):88–115

Park SY, Lee JH, Ha M, Nam JW, Kim VN (2009) miR-29 miRNAs activate p53 by targeting p85 alpha and CDC42. Nat Struct Mol Biol 16(1):23–29

Passtoors WM, Beekman M, Deelen J, van der Breggen R, Maier AB, Guigas B, Derhovanessian E, van Heemst D, de Craen AJ, Gunn DA, Pawelec G, Slagboom PE (2013) Gene expression analysis of mTOR pathway: association with human longevity. Aging Cell 12(1):24–31

Peleg S, Sananbenesi F, Zovoilis A, Burkhardt S, Bahari-Javan S, Agis-Balboa RC, Cota P, Wittnam JL, Gogol-Doering A, Opitz L, Salinas-Riester G, Dettenhofer M, Kang H, Farinelli L, Chen W, Fischer A (2010) Altered histone acetylation is associated with age-dependent memory impairment in mice. Science 328(5979):753–756

Perry JJ, Shin DS, Getzoff ED, Tainer JA (2010) The structural biochemistry of the superoxide dismutases. Biochim Biophys Acta 1804(2):245–262

Pirooznia SK, Elefant F (2013) Targeting specific HATs for neurodegenerative disease treatment: translating basic biology to therapeutic possibilities. Front Cell Neurosci 7:30

Poon HF, Vaishnav RA, Getchell TV, Getchell ML, Butterfield DA (2006) Quantitative proteomics analysis of differential protein expression and oxidative modification of specific proteins in the brains of old mice. Neurobiol Aging 27(7):1010–1019

de Pril R, Fischer DF, Maat-Schieman ML, Hobo B, de Vos RA, Brunt ER, Hol EM, Roos RA, van Leeuwen FW (2004) Accumulation of aberrant ubiquitin induces aggregate formation and cell death in polyglutamine diseases. Hum Mol Genet 13(16):1803–1813

Qiu J (2006) Epigenetics: unfinished symphony. Nature 441(7090):143–145

Ran Q, Liang H, Ikeno Y, Qi W, Prolla TA, Roberts LJ 2nd, Wolf N, Van Remmen H, Richardson A (2007) Reduction in glutathione peroxidase 4 increases life span through increased sensitivity to apoptosis. J Gerontol A Biol Sci Med Sci 62(9):932–942

Rao KS (2007) DNA repair in aging rat neurons. Neuroscience 145(4):1330–1340

Razzaque MS (2012) The role of Klotho in energy metabolism. Nat Rev Endocrinol 8(10):579–587

Romanov GA, Vanyushin BF (1981) Methylation of reiterated sequences in mammalian DNAs. Effects of the tissue type, age, malignancy and hormonal induction. Biochim Biophys Acta 653(2):204–218

Roth GS, Ingram DK, Joseph JA (2007) Nutritional interventions in aging and age-associated diseases. Ann N Y Acad Sci 1114:369–371

Salih DA, Brunet A (2008) FoxO transcription factors in the maintenance of cellular homeostasis during aging. Curr Opin Cell Biol 20(2):126–136

Schindeldecker M, Stark M, Behl C, Moosmann B (2011) Differential cysteine depletion in respiratory chain complexes enables the distinction of longevity from aerobicity. Mech Ageing Dev 132(4):171–179

Schmidt U, Holsboer F, Rein T (2011) Epigenetic aspects of posttraumatic stress disorder. Dis Markers 30(2–3):77–87

Sebastiani P, Solovieff N, Dewan AT, Walsh KM, Puca A, Hartley SW, Melista E, Andersen S, Dworkis DA, Wilk JB, Myers RH, Steinberg MH, Montano M, Baldwin CT, Hoh J, Perls TT (2012) Genetic signatures of exceptional longevity in humans. PLoS One 7(1):e29848

Seluanov A, Chen Z, Hine C, Sasahara TH, Ribeiro AA, Catania KC, Presgraves DC, Gorbunova V (2007) Telomerase activity coevolves with body mass not lifespan. Aging Cell 6(1):45–52

Shay JW, Wright WE (2007) Hallmarks of telomeres in ageing research. J Pathol 211(2):114–123

Shay T, Jojic V, Zuk O, Rothamel K, Puyraimond-Zemmour D, Feng T, Wakamatsu E, Benoist C, Koller D, Regev A, ImmGen Consortium (2013) Conservation and divergence in the transcriptional programs of the human and mouse immune systems. Proc Natl Acad Sci U S A 110(8):2946–2951

Shumaker DK, Dechat T, Kohlmaier A, Adam SA, Bozovsky MR, Erdos MR, Eriksson M, Goldman AE, Khuon S, Collins FS, Jenuwein T, Goldman RD (2006) Mutant nuclear lamin A leads to progressive alterations of epigenetic control in premature aging. Proc Natl Acad Sci U S A 103(23):8703–8708

Sies H (1986) Biochemistry of oxidative stress. Angewandte Chemie Int Ed 12:1058–1071

Soto C, Estrada LD (2008) Protein misfolding and neurodegeneration. Arch Neurol 65(2):184–189

Squier TC (2001) Oxidative stress and protein aggregation during biological aging. Exp Gerontol 36(9):1539–1550

Stadtman ER (2006) Protein oxidation and aging. Free Radic Res 40(12):1250–1258

Steves CJ, Spector TD, Jackson SH (2012) Ageing, genes, environment and epigenetics: what twin studies tell us now, and in the future. Age Ageing 41(5):581–586

Strong R, Miller RA, Astle CM, Floyd RA, Flurkey K, Hensley KL, Javors MA, Leeuwenburgh C, Nelson JF, Ongini E, Nadon NL, Warner HR, Harrison DE (2008) Nordihydroguaiaretic acid and aspirin increase lifespan of genetically heterogeneous male mice. Aging Cell 7(5):641–650

Surova O, Zhivotovsky B (2013) Various modes of cell death induced by DNA damage. Oncogene 32(33):3789–3797

Sykora P, Wilson DM 3rd, Bohr VA (2013) Base excision repair in the mammalian brain: implication for age related neurodegeneration. Mech Ageing Dev 134(10):440–448

Szilard L (1959) On the nature of the aging process. Proc Natl Acad Sci U S A 45(1):30–45

Tam JH, Pasternak SH (2012) Amyloid and Alzheimer's disease: inside and out. Can J Neurol Sci 39(3):286–298

Tammen SA, Friso S, Choi SW (2013) Epigenetics: the link between nature and nurture. Mol Aspects Med 34(4):753–764

Tan Y, Bush JM, Liu W, Tang F (2009) Identification of longevity genes with systems biology approaches. Adv Appl Bioinform Chem 2:49–56

Tatar M, Khazaeli AA, Curtsinger JW (1997) Chaperoning extended life. Nature 390:30

Tatar M, Kopelman A, Epstein D, Tu MP, Yin CM, Garofalo RS (2001) A mutant Drosophila insulin receptor homolog that extends life-span and impairs neuroendocrine function. Science 292(5514):107–110

Tomás-Loba A, Flores I, Fernández-Marcos PJ, Cayuela ML, Maraver A, Tejera A, Borrás C, Matheu A, Klatt P, Flores JM, Viña J, Serrano M, Blasco MA (2008) Telomerase reverse transcriptase delays aging in cancer-resistant mice. Cell 135(4):609–622

Vallabhaneni H, O'Callaghan N, Sidorova J, Liu Y (2013) Defective repair of oxidative base lesions by the DNA glycosylase Nth1 associates with multiple telomere defects. PLoS Genet 9(7):e1003639

Van Raamsdonk JM, Hekimi S (2012) Superoxide dismutase is dispensable for normal animal lifespan. Proc Natl Acad Sci U S A 109(15):5785–5790

Vanyushin BF, Nemirovsky LE, Klimenko VV, Vasiliev VK, Belozersky AN (1973b) The 5-methylcytosine in DNA of rats. Tissue and age specificity and the changes induced by hydrocortisone and other agents. Gerontologia 19(3):138–152

Vanyushin BF, Mazin AL, Vasilyev VK, Belozersky AN (1973a) The content of 5-methylcytosine in animal DNA: the species and tissue specificity. Biochim Biophys Acta 299(3):397–403

Vessoni AT, Filippi-Chiela EC, Menck CF, Lenz G (2013) Autophagy and genomic integrity. Cell Death Differ 20(11):1444–1454

Vilenchik MM, Knudson AG Jr (2000) Inverse radiation dose-rate effects on somatic and germ-line mutations and DNA damage rates. Proc Natl Acad Sci U S A 97(10):5381–5386

Villalba JM, Alcaín FJ (2012) Sirtuin activators and inhibitors. Biofactors 38(5):349–359

Villalba JM, de Cabo R, Alcain FJ (2012) A patent review of sirtuin activators: an update. Expert Opin Ther Pat 22(4):355–367

Vyjayanti VN, Rao KS (2006) DNA double strand break repair in brain: reduced NHEJ activity in aging rat neurons. Neurosci Lett 393(1):18–22

Waddington CH (2012) The epigenotype. 1942. Int J Epidemiol 41(1):10–13

Weiss EP, Fontana L (2011) Caloric restriction: powerful protection for the aging heart and vasculature. Am J Physiol Heart Circ Physiol 301(4):H1205–H1219

Wilkinson KD, Urban MK, Haas AL (1980) Ubiquitin is the ATP-dependent proteolysis factor I of rabbit reticulocytes. J Biol Chem 255:7529–7532

Witte AV, Fobker M, Gellner R, Knecht S, Flöel A (2009) Caloric restriction improves memory in elderly humans. Proc Natl Acad Sci U S A 106(4):1255–1260

Wong AS, Cheung ZH (1812) Ip NY (2011) Molecular machinery of macroautophagy and its deregulation in diseases. Biochim Biophys Acta 11:1490–1497

Xiong N, Long X, Xiong J, Jia M, Chen C, Huang J, Ghoorah D, Kong X, Lin Z, Wang T (2012) Mitochondrial complex I inhibitor rotenone-induced toxicity and its potential mechanisms in Parkinson's disease models. Crit Rev Toxicol 42(7):613–632

Xu G, Herzig M, Rotrekl V, Walter CA (2008) Base excision repair, aging and health span. Mech Ageing Dev 129(7–8):366–382

Yakar S, Adamo ML (2012) Insulin-like growth factor 1 physiology: lessons from mouse models. Endocrinol Metab Clin North Am 41(2):231–247

Yang Z, Klionsky DJ (2010) Mammalian autophagy: core molecular machinery and signaling regulation. Curr Opin Cell Biol 22:124–131

Yi C, He C (2013) DNA repair by reversal of DNA damage. Cold Spring Harb Perspect Biol 5(1):a012575

Yin F, Jiang T, Cadenas E (2013) Metabolic triad in brain aging: mitochondria, insulin/IGF-1 signalling and JNK signalling. Biochem Soc Trans 41(1):101–105

Young JC (2010) Mechanisms of the Hsp70 chaperone system. Biochem Cell Biol 88(2):291–300

Zentner GE, Henikoff S (2013) Regulation of nucleosome dynamics by histone modifications. Nat Struct Mol Biol 20(3):259–266

Zschocke J, Manthey D, Bayatti N, van der Burg B, Goodenough S, Behl C (2002) Estrogen receptor alpha-mediated silencing of caveolin gene expression in neuronal cells. J Biol Chem 277(41):38772–38780

Zuckerman V, Wolyniec K, Sionov RV, Haupt S, Haupt Y (2009) Tumour suppression by p53: the importance of apoptosis and cellular senescence. J Pathol 219(1):3–15

Chapter 4
Selected Age-Related Disorders

Abstract Aged human individuals are frequently affected by various age-related impairments and disorders at once (multimorbidity) making it difficult to investigate and understand the link between aging as key risk factor and single syndromes. In some chapters of this book links have been identified between molecular mechanisms of aging and the pathogenesis of human disorders in the elderly. Here, the focus is put on Alzheimers disease (AD) and cancer. While cancer in many cases can successfully be treated and sometimes even cured by pharmacology and/or surgery, AD is still an incurable deadly disease. Like almost no other disorder the onset of AD is strictly associated with higher age and experiences a strong increase in case numbers in our aging society. The authors are convinced that the actual causes of AD can only be identified when the biochemistry of the aging neuron is unraveled, especially in the context of well-known life-time risk factors.

Keywords Age-related disorders · Neurodegeneration · Brain aging · Alzheimers disease · Cancer · Age metabolism

The implicature of the terms "age" and "aging" is mostly negative which probably is due to the fact that it is hard to accept that there is a progressive decline of certain body functions and other obvious age signs including skin wrinkling, age-spots or grey hair. While humans may get along with general age-related changes and may even effectively adapt and partly counteract unwanted alterations there is a large number of conditions of serious age-associated functional decline and disorders (see also Figs. 1.2 and 1.3). Aging is accepted as one of the biggest risk factors for the development of many diseases including neurodegenerative disorders, cancer and type-2 diabetes. These disorders are believed to be a direct result of a combination of various genetic, lifestyle, and environmental factors (Fransen et al. 2013; López-Otín et al. 2013). It is far beyond the scope of this book to present and fully discuss these disorders that more frequently occur in the elderly with age being the main risk factor. And, of course, a subset of such pathologies occur also earlier in life depending to some part on genetic predisposition as, for instance, certain types of cancer. The fact that aged individuals frequently experience multiple disorders, called multimorbidity and creating a snowball effect of health conditions makes an understanding of aging as key risk factor of certain age-related disorders complicated. As we slowly begin to understand, an aged cell is different in its biochemistry and physiology when compared to a young one. Consequently, the aging process itself precipitates many changes in cell, tissue and body function. The range of severity of the latter is wide and reaches from being inconvenient to highly cumbersome. But even mild negative changes in body function can increase the severity of other age-associated pathological conditions. The most frequently reported and observed age-associated changes in body function concern the general metabolism (which may lead to increased body fat), liver and kidney function, sleep behavior, and menopause in the female organism. In general, the following defined disorders occur in the elderly population at an increased risk: osteoporosis, cardiovascular disorders mostly based on age-associated arteriosclerosis and hypertension, diabetes type II, osteoarthritis associated with chronic inflammation and pain, certain types of cancer, and neurodegeneration, namely Alzheimers disease. In addition to this progressive chronic conditions medical statistics teach us that also several acute health conditions occur more frequently with increasing age, such as cardiac infarction, heart attack or stroke. The reason for this increase is obvious, namely the chronic pathological conditions and impaired or loss of body functions mentioned above. In the following two completely different but highly frequent and age-associated types of disorders should be presented in some more detail but not with respect to suggested pathogenetic theories, cause and therapies but only in the context of their relationship to aging. The focus will be on Alzheimers disease and cancer since age is widely accepted as the main risk factor for the most prevalent disorders in developed countries, including neurodegeneration, cancer as well as cardiovascular disease which has not been addressed here in detail (Niccoli and Partridge 2012).

4.1 Age: The Key Risk Factor for Alzheimers Disease

"Choose your parents carefully and die young" was the reply of the famous US Alzheimer expert Dennis Selkoe from Harvard when asked how to avoid getting Alzheimers disease. AD is still an incurable, progressive, deadly disease of the brain and a huge medical but also socioeconomic and ethical challenge and burden for our aging society. As simple as Selkoe's statement appears it summarized what we definitively know about the development of AD up to date, the powerful role of certain genes for a minority of AD cases and the key influence of the human age for the majority. There are certain gene mutations that definitively will lead to AD, frequently already at young age. But this familial AD cases only account for less than 5 % of all AD patients. By far the majority of AD cases are strictly age-related (sporadic forms) while due to limited diagnostic possibilities to define AD *pre mortem* the exact numbers vary. But is clear that there is a strict correlation between AD onset and human age. Estimates range up to a frequency of 20–25 % when people reach the age of 85 and higher (Ballard et al. 2011). With age as the central risk factor for

AD, dying young indeed "prevents" age-associated sporadic AD. The fact that life expectancy has increased so much in the last 100 years consequently also leads to a much higher number of age-associated AD. Despite the general acknowledgement of aging as AD risk factor the relationship between aging and AD still remains unresolved. In the following, normal aging of the brain, the devolpment of dementia and AD will be discussed in more detail.

Normal aging and the sporadic AD brain: There is a continuous decline in general brain function with age which is described to start as early as in the 20s. One key biological substrate for such a decline are, for instance, observed changes in synaptic plasticity (Bartrés-Faz and Arenaza-Urquijo 2011; DeCarli et al. 2012). Therefore, based on neuropsychological testing the following brain functions decrease with brain aging, namely speed of processing, working memory, long-term memory, fluid intelligence, explicit memory (episodic prior to semantic), alertness with mainly vigilance and flexibility and general problem solving. On the other hand some brain functions are improved during lifetime including verbal knowledge. It is of general acceptance that learning is harder in older age and that the brain faces reduced short-term memory capabilities. In contrast, early aquired memory and life history remains present (e.g. Erraji-Benchekroun et al. 2005; Caserta et al. 2009; Wagster 2009; Park and Bischof 2013). Anatomically functional impairment is ascribed to changes in the prefrontal cortex (PFC) and the hippocampus (Jellinger and Attems 2013). The PFC is the anterior part of the frontal lobes of the brain and has been strongly implicated in planning complex cognitive behavior, personality expression, decision making, and moderating social behavior. The hippocampus is a major component of the so-called limbic system and has been shown to play important roles in the general consolidation of information from short-term memory to long-term memory as well as spatial navigation (Kandel 2001; Kandel et al. 2012). Emplyoing microarray technology the *post mortem* PFC material of 39 human subjects with the age ranging from 13 to 79 years was analyzed. In summary, this study found that there is life-long progressive change in gene expression with age but only in approx. 7.5 % of genes of the human genome. Consequently, expression levels for the large majority of genes were strikingly unaltered throughout adult life. It was concluded that a small set of genes in specific cellular populations and biological processes are selectively struck during aging (Erraji-Benchekroun et al. 2005). The affected genes could be grouped in genes associated with glial-mediated inflammation, the oxidative stress response, mitochondrial function, synaptic function and plasticity and calcium regulation. Overall it was concluded that genes that are upregulated are in most cases of glial origin and related to inflammation and cellular defense pathways. On the other hand, genes that are downregulated during aging display mostly neuron-enriched transcripts that can be associated with cellular communication and signaling. The authors even propose that because these age-related changes are highly consistent and specific they may serve as biomarkers of "molecular age" that can be used to predict the indivdual age, based on gene expression profile (Erraji-Benchekroun et al. 2005; Sibille 2013).

In contrast to AD brains, in healthy aged brains there is no reduction in neuron number . Consequently, neurodegeneration and neuronal cell loss in AD are due to

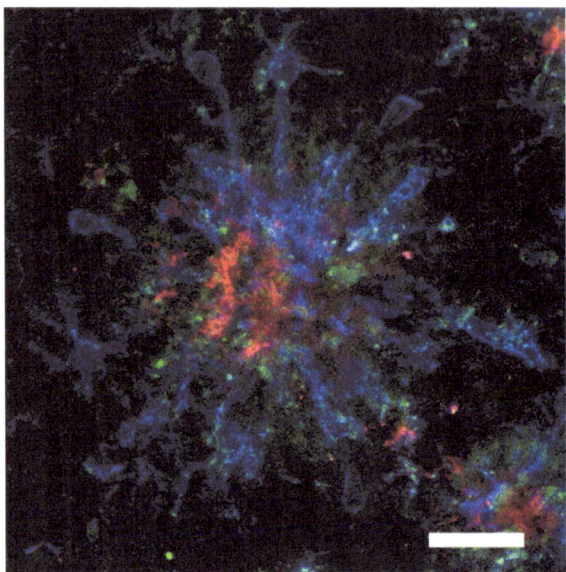

Fig. 4.1 Amyloid plaques surrounded by microglia cells. Section of hippocampus of an APP23 mouse, an animal model for Alzheimers disease overexpressing a mutated form of amyloid precursor protein (APP). (Immuno)fluorescence staining of amyloid plaques (anti-amyloid beta, *red colour*) and activated microglia cells (tomatolectin, *green* colour and anti-Iba I, *blue colour*). Scale bar: 20 μM (photo courtesy of Christof Hiebel, Institute for Pathobiochemistry, University Medical Center Mainz)

a distinct pathological process (Burke and Barnes 2006). The expansion of brain atrophy in different brain regions directly correlates with the cognitive decline in sporadic AD (Scahill et al. 2002; Jack Jr et al. 2004). One AD neuropathological hallmark are neurofibrillary tangles (NFTs), a special protein structure made up of a chemically modified tau-protein involved in protein transport in the neuron, in different regions of the brain. Interestingly, non-demented elderly frequently also show significant levels of NFTs. And when analyzing the brain of non AD people at an age of 85 almost everyone studied will display such NFTs in the cerebral cortex (Ohm et al. 1995). In AD the progression of neurodegeneration directly correlates with the total number of NFTs and the particular site they are located in the brain (Giannakopoulos et al. 2003). Therefore, NFTs are considered as early factor of AD and their occurence is widely used for *post mortem* staging of cognitive deficits (Braak and Braak 1991). A second main hallmark defining the AD histopathology is the extracellular deposition of the amyloid β protein (Aβ; Fig. 4.1), a cleavage product of a larger precursor protein (amyloid precursor protein, APP). In contrast to NFTs no undoubted correlation of the tissue load of Aβ and the severity of cognitive deficits has been found. In fact, rather contradictory data were published on that important issue but the view that AD-associated cognitive deficits and memory impairment are independent of the extent of Aβ deposition (in so called senile plaques) is getting more and more acknowledged and recent findings also from the authors lab significantly fuel this discussion (Veeraraghavalu et al. 2013; Price et al. 2013; Stumm et al. 2013). Moreover, Aβ levels have been described to be increased in cognitively healthy elderly and a substantial amount of Aβ plaques can be observed (Schupf et al. 2008). Taking together the histological data, general alterations observed in the brain during healthy aging are detectable but not dramatic and can include the presence of considerable levels of NFTs and Aβ containing plaques suggesting that the occurence of Aβ and NFTs alone, used as histopathological markers for *post mortem* diagnosis of AD, is not sufficient to explain AD-associated neuronal loss and gross brain atrophy. Obviously, additional age-related factors are essential to initiate AD-associated neurodegenerative events.

Since AD is characterized by a loss in cholinergic transmission different studies have focused on changes in cholinergic neurons. Acetylcholine is one of the essential excitatory neurotransmitters in the central nervous system (Kern and Behl 2009; Ballard et al. 2011). Neurotransmission, especially by acetylcholine and the amino acid glutamate are key targets of the current pharmacological intervention aiming to improve cognitive function in AD patients (Pohanka 2012; Corbett et al. 2012) as neurotransmission is significantly altered and more rapidly deteriorated in sporadic AD. Various general age-related risk factors that directly affect the development of sporadic AD are known demonstrating an actual interplay between aging and AD. Epidemiology demonstrates hypertension, hypercholesterolemia, obesity, diabetes and inflammation as strong conditions that influence onset and progression of sporadic AD. For instance, with respect to hypertension a positive correlation between high blood pressure in midlife and the onset of cognitive impairment later in life has been described (Knopman et al. 2001). Chronic hypertension can cause vascular lesions promoting the onset of cognitive decline. The prominent age-related cardiovascular risk factors such as high cholesterol (hypercholesterolemia), obesity and diabetes are all based on metabolic disturbances and are as such strongly affecting the prevalence of cognitive decline and potentially AD (Vanhanen et al. 2006; Whitmer et al. 2005). Another age-related potential disposition factor for AD is inflammation and it has been shown that the activation of immune cells in the brain (e.g. astrocytes) is constantly increasing with age (Sastre et al. 2006). There are several links of astrocyte function and Aβ biochemistry and in AD brains there is increased gliosis surrounding Aβ plaques compared to non-demented controls (Vehmas et al. 2003).

An additional age risk factor for sporadic AD is oxidative stress caused by reactive oxygen and nitrogen species which were already discussed in the context of the free radical theory of aging by Harman (1956). After decades of research and a load of experimental data also an *oxidative stress theory of neurodegeneration* (including AD) can be formulated clearly demonstrating how aging and neurodegeneration may funnel into the same biochemical cascade of events. And, indeed, an increased oxidative burden mainly shown by the detection of oxidized biomolecules (DNA, proteins, lipids) in brain tissue has been described for the brain of non-demented elderly as well as of sporadic AD patients (Moosmann and Behl 2002; Keller et al. 2005; Zhu et al. 2006). Interestingly, it is also well-established that mutations in the genome of mitochondria, the main sources of free oxygen radicals, accumulate during brain aging as well as in neurodegenerative diseases (Corral-Debrinski et al. 1992; Bender et al. 2006) showing that these two processes potentially share common pathological pathways. Taken many experimental studies and histopathological observations together it can be summarized that the accumulation of oxidatively modified biomolecules is a central hallmark of brain aging and is enhanced during neurodegenerative conditions such as AD. Finally it can be acknowledged that alterations occuring in the aging brain at various levels do also occur in sporadic AD and age-associated general changes represent direct predisposition conditions and can increase the incidence of AD. Consequently, a general improvement in health of the elderly person by minimizing risk factors, for instance by special diets and changes in general lifestyle parameters, may reduce the probability getting age-associated AD or at least postpone its onset to a higher age. So, brain aging and AD are intimately related to each other.

4.2 The Common Biology of Aging and Cancer: Senescence as Cancer Protection

Although cancer can occur throughout the whole life (partially depending on the individual genetic background), there is a steep increase in cancer incidence during aging (Crawford and Cohen 1987), clearly demonstrated, for instance, by the higher incidence of prostate cancer in men and breast cancer in women. Consequently, cancer can also be seen as age-associated pathology and in certain aspects evidence of a common biology of cancer and aging is growing. There are various mechanistical overlaps and knowledge out of the cancer field that has been translated into aging biology. In particlular five main aspects are ususally addressed in this context and have been excellently reviewed in detail including (1) the direct link between cellular senescence and tumor formation, (2) the role of genome stability and instability, (3) the role of telomers, (4) the significance of autophagy in cancer and aging, and (5) the role of mitochondrial energy balance (Finkel et al. 2007; Vijg and Suh 2013; Pereira and Ferreira 2013). Obviously, there is a thin line between cells and tissues that are aging and, ultimately, being removed via apoptosis and those that are switching into a constantly proliferating state. These processes call for a tight control and so aging (senescence followed by subsequent apoptosis) means cancer protection. Transferring cells into a senescence state limits the proliferative potential of damaged cells.

With respect to the maintenance of genome stability it is of interest that DNA damage can lead to cancer development but can also shift a cell into senescence or directly into apoptosis (see Fig. 3.7). As it was discussed earlier, genomic instability is a key hallmark of aging and DNA damage when escaping the cell cycle control checkpoints can induce DNA mutations that may lead to dysfunction or uncontrolled cell proliferation. Aging and cancer are also connected via the pathway of autophagy. Autophagy whose link to aging has been described in some more detail

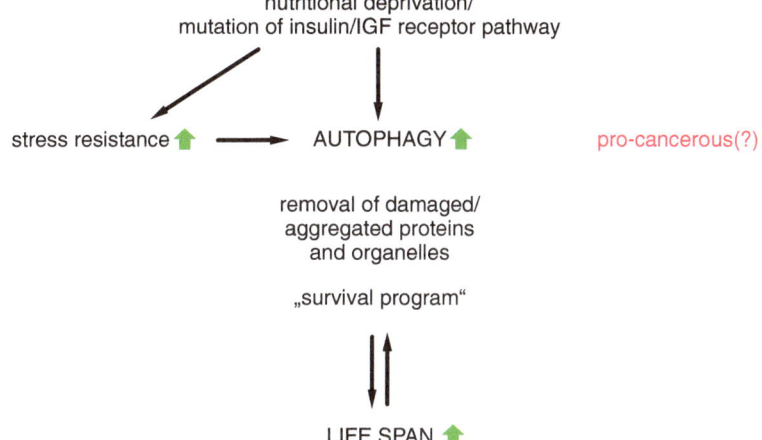

Fig. 4.2 Autophagy is connected to cancer and aging. Autophagy is an evolutionary conserved degradation process and a central survival mechanism when cells are under unfavorable conditions or stress. On the other hand cells can also escape exogenous challenges and cell cycle control by an upregulation of autophagy which gives autophagy also a potential pro-cancerous function

above helps cells to survive under reduced nutrient or stress conditions. Therefore, autophagy can be considered as being pro-cancerous in that it enables cancer cells to survive even under low oxygen (hypoxia) or upon challenge with radiation or chemotherapy (exogenous stress). Consequently, autophagy is a reasonable target for cancer therapy although it is hard to aim at since autophagy is such a multi-controlled complex and basic mechanism. In addition, a pathway called *autophagic cell death* triggered by tumor suppressors (p53) has been defined which is supposed to drive cells into death similar to apoptosis (Crighton et al. 2006). But recently, this view has been challenged and, again, is hard to analyze experimentally due to the complexity of this process; autophagy may also function as tumor suppressor as recently discussed in detail (Shen et al. 2012; Giuliani and Dass 2013). Whatever function dominates is depending on the individual situation the cell is in, the nutritional status, whether the cell is under oxidative or genotoxic stress etc. (Vessoni et al. 2013; Fig. 4.2). With respect to the link of autophagy with aging key models of life span manipulation in *C.elegans* were already presented. In the daf-2-deficient worm line exhibiting an extended life span the insulin/IGF receptor is removed mimicking starvation conditions. Such a nutritional deprivation is also a potent inducer of autophagy in *C.elegans*. So here age-related processes and autophagy directly merge. Interestingly, in mammalian model tissues genetic inhibition of autophagy induces degenerative alterations that are comparable to those seen during aging. Moreover, physiological (normal) and non-physiological (pathological) aging are often associated with a reduced autophagy. Besides, pharmacological and as just mentioned genetic manipulations leading to life span extension in model organisms often induce and stimulate autophagy (Rubinsztein et al. 2011).

This directly leads to another connection between aging and cancer which concerns the general metabolism of the cells. The manipulation of pathways involved in the utilization of nutrition, such as the insulin/IGF-receptor pathway in *C.elegans*, leads to an attenuation of aging and a prolonged life span. Frequently, this is also combined with an increased stress resistance that is also a promoting factor for tumor growth and cancer development. Moreover, in a clonal hippocampal tumor cell line the induction of resistance against oxidatve stress leads to stable and rapidly growing clonal variants that show a massive increase in lysosome formation and autophagy markers (Clement et al. 2009, 2010).

Since there are so many overlaps between cancer and aging on one hand, on the other hand it is highly relevant to uncover the mechanisms that guide a cell in either one direction. The example that DNA damage can induce aging as well as cancer is a paradigm where this control can be investigated. Eventually the ultimate decision point occurs to keep a proliferative state (which bears the danger of tumor formation) or to release cells into aging (as an exit strategy), feeding them into the senescence state and, ultimately, apoptosis. The latter means to sort out single cells and tissue that may potentially harm the whole organism. Therefore, mechanisms and players that influence individual cell aging, such as telomers, may serve to maintain a healthy and functional life as long as possible and necessary for the function of the organ. Aging of the cell as then initiated by telomer shortening controls the proliferation of cells and reduces the probability of the development of cancer. Therefore, aging would not be a process occuring eventually and randomly induced by external factors but would rather have the goal to counteract an early death of the whole organism through unlimited proliferation and cancer. In other words, aging is the price we have to pay for surviving.

Finally, "aging seems to be the only available way to live a long life."-Daniel Francois Auber: If we could have one wish for free it might the one to live a long and healthy life which then ends instantly and without any suffering or the need for long time care with low life quality. Research into the understanding of aging should go into exactly this direction rather than in trying to find ways that allow unlimited life span and immortality, because after all "Who wants to live forever?" (Queen/Brian May, Soundtrack for *Highlander* 1986).

References

Auber Daniel Francois Esprit (1997) In: Bloomsbury biographic dictionary of quotations. Bloomsbury, London

Ballard C, Gauthier S, Corbett A, Brayne C, Aarsland D, Jones E (2011) Alzheimer's disease. Lancet 377(9770):1019–1031

Bartrés-Faz D, Arenaza-Urquijo EM (2011) Structural and functional imaging correlates of cognitive and brain reserve hypotheses in healthy and pathological aging. Brain Topogr 24(3–4): 340–357

Bender A, Krishnan KJ, Morris CM, Taylor GA, Reeve AK, Perry RH, Jaros E, Hersheson JS, Betts J, Klopstock T, Taylor RW, Turnbull DM (2006) High levels of mitochondrial DNA deletions in substantia nigra neurons in aging and Parkinson disease. Nat Genet 38(5):515–517

Braak H, Braak E (1991) Neuropathological stageing of Alzheimer-related changes. Acta Neuropathol 82(4):239–259

Burke SN, Barnes CA (2006) Neural plasticity in the ageing brain. Nat Rev Neurosci 7:30–40

Caserta MT, Bannon Y, Fernandez F, Giunta B, Schoenberg MR, Tan J (2009) Normal brain aging clinical, immunological, neuropsychological, and neuroimaging features. Int Rev Neurobiol 84:1–19

Clement AB, Gamerdinger M, Tamboli IY, Lütjohann D, Walter J, Greeve I, Gimpl G, Behl C (2009) Adaptation of neuronal cells to chronic oxidative stress is associated with altered cholesterol and sphingolipid homeostasis and lysosomal function. J Neurochem. 111(3):669–682

Clement AB, Gimpl G, Behl C (2010) Oxidative stress resistance in hippocampal cells is associated with altered membrane fluidity and enhanced nonamyloidogenic cleavage of endogenous amyloid precursor protein. Free Radic Biol Med 48(9):1236–1241

Corbett A, Smith J, Ballard C (2012) New and emerging treatments for Alzheimer's disease. Expert Rev Neurother 12(5):535–543

Corral-Debrinski M, Horton T, Lott MT, Shoffner JM, Beal MF, Wallace DC (1992) Mitochondrial DNA deletions in human brain: regional variability and increase with advanced age. Nat Genet 2:324–329

Crawford J, Cohen HJ (1987) Relationship of cancer and aging. Clin Geriatr Med 3(3):419–432

Crighton D, Wilkinson S, O'Prey J, Syed N, Smith P, Harrison PR, Gasco M, Garrone O, Crook T, Ryan KM (2006) DRAM, a p53-induced modulator of autophagy, is critical for apoptosis. Cell 126(1):121–134

DeCarli C, Kawas C, Morrison JH, Reuter-Lorenz PA, Sperling RA, Wright CB (2012) Session II: Mechanisms of age-related cognitive change and targets for intervention: neural circuits, networks, and plasticity. J Gerontol A Biol Sci Med Sci 67(7):747–753

Erraji-Benchekroun L, Underwood MD, Arango V, Galfalvy H, Pavlidis P, Smyrniotopoulos P, Mann JJ, Sibille E (2005) Molecular aging in human prefrontal cortex is selective and continuous throughout adult life. Biol Psychiatry 57(5):549–558

Finkel T, Serrano M, Blasco MA (2007) The common biology of cancer and ageing. Nature 448(7155):767–774

Fransen M, Nordgren M, Wang B, Apanasets O, Van Veldhoven PP (2013) Aging, age-related diseases and peroxisomes. Subcell Biochem 69:45–65

Giannakopoulos P, Herrmann FR, Bussière T, Bouras C, Kövari E, Perl DP, Morrison JH, Giuliani CM, Dass CR (2013) Autophagy and cancer: taking the 'toxic' out of cytotoxics. J Pharm Pharmacol 65(6):777–789

Giuliani CM, Dass CR (2013) Autophagy and cancer: taking the 'toxic' out of cytotoxics. J Pharm Pharmacol 65(6):777–789

Gold G, Hof PR (2003) Tangle and neuron numbers, but not amyloid load, predict cognitive status in Alzheimer's disease. Neurology 60:1495–1500

Harman D (1956) Aging: a theory based on free radical and radiation chemistry. J Gerontol 11(3):298–300

Jack CR Jr, Shiung MM, Gunter JL, O'Brien PC, Weigand SD, Knopman DS, Boeve BF, Ivnik RJ, Smith GE, Cha RH, Tangalos EG, Petersen RC (2004) Comparison of different MRI brain atrophy rate measures with clinical disease progression in AD. Neurology 62:591–600

Jellinger KA, Attems J (2013) Neuropathological approaches to cerebral aging and neuroplasticity. Dialogues Clin Neurosci 15(1):29–43

Kandel ER (2001) The molecular biology of memory storage: a dialogue between genes and synapses. Science 294(5544):1030–1038

Kandel ER, Schwartz JH, Jessell TM, Siegelbaum SA, A. J. Hudspeth AJ (2012) Principles of Neural Science, 5th Edn. McGraw-Hill Professional, New York

Keller JN, Schmitt FA, Scheff SW, Ding Q, Chen Q, Butterfield DA, Markesbery WR (2005) Evidence of increased oxidative damage in subjects with mild cognitive impairment. Neurology 64:1152–1156

Kern A, Behl C (2009) The unsolved relationship of brain aging and late-onset Alzheimer disease. Biochim Biophys Acta 1790(10):1124–1132

Neurology (2001) Cardiovascular risk factors and cognitive decline in middle-aged adults. 56:42–48

López-Otín C, Blasco MA, Partridge L, Serrano M, Kroemer G (2013) The hallmarks of aging. Cell 153(6):1194–1217

Moosmann B, Behl C (2002) Antioxidants as treatment for neurodegenerative disorders. Expert Opin Investig Drugs 11(10):1407–1435

Niccoli T, Partridge L (2012) Ageing as a risk factor for disease. Curr Biol 22(17):R741–752

Ohm TG, Müller H, Braak H, Bohl J (1995) Close-meshed prevalence rates of different stages as a tool to uncover the rate of Alzheimer's disease-related neurofibrillary changes. Neuroscience 64(1):209–217

Park DC, Bischof GN (2013) The aging mind: neuroplasticity in response to cognitive training. Dialogues Clin Neurosci 15(1):109–119

Pereira B, Ferreira MG (2013) Sowing the seeds of cancer: telomeres and age-associated tumorigenesis. Curr Opin Oncol 25(1):93–98

Pohanka M (2012) Acetylcholinesterase inhibitors: a patent review (2008 - present). Expert Opin Ther Pat 22(8):871–886

Price AR, Xu G, Siemienski ZB, Smithson LA, Borchelt DR, Golde TE, Felsenstein KM (2013), Comment on "ApoE-directed therapeutics rapidly clear β-amyloid and reverse deficits in AD mouse models". Science 340(6135):924–d

Rubinsztein DC, Mariño G, Kroemer G (2011) Autophagy and aging. Cell 146(5):682–695

Sastre M, Klockgether T, Heneka MT (2006) Contribution of inflammatory processes to Alzheimer's disease: molecular mechanisms. Int J Dev Neurosci 24:167–176

Scahill RI, Schott JM, Stevens JM, Rossor MN, Fox NC (2002) Mapping the evolution of regional atrophy in Alzheimer's disease: unbiased analysis of fluid-registered serial MRI. Proc Natl Acad Sci USA 99:4703–4707

Schupf N, Tang MX, Fukuyama H, Manly J, Andrews H, Mehta P, Ravetch J, Mayeux R (2008) Peripheral Abeta subspecies as risk biomarkers of Alzheimer's disease. Proc Natl Acad Sci USA 105:14052–14057

Shen S, Kepp O, Kroemer G (2012) The end of autophagic cell death? Autophagy 8(1):1–3

Sibille E (2013) Molecular aging of the brain, neuroplasticity, and vulnerability to depression and other brain-related disorders. Dialogues Clin Neurosci 15(1):53–65

Stumm C, Hiebel C, Hanstein R, Purrio M, Nagel H, Conrad A, Lutz B, Behl C, Clement AB (2013) Cannabinoid receptor 1 deficiency in a mouse model of Alzheimer's disease leads to enhanced cognitive impairment despite a reduction in amyloid deposition. Neurobiol Aging 34(11):2574–2584

Vanhanen M, Koivisto K, Moilanen L, Helkala EL, Hänninen T, Soininen H, Kervinen K, Kesäniemi YA, Laakso M, Kuusisto J (2006) Association of metabolic syndrome with Alzheimer disease: a population-based study. Neurology 67:843–847

Veeraraghavalu K, Zhang C, Miller S, Hefendehl JK, Rajapaksha TW, Ulrich J, Jucker M, Holtzman DM, Tanzi RE, Vassar R, Sisodia SS (2013), Comment on "ApoE-directed therapeutics rapidly clear β-amyloid and reverse deficits in AD mouse models". Science 340(6135):924–f

Vehmas AK, Kawas CH, Stewart WF, Troncoso JC (2003) Immune reactive cells in senile plaques and cognitive decline in Alzheimer's disease. Neurobiol Aging 24:321–331

Vessoni AT, Filippi-Chiela EC, Menck CF, Lenz G (2013) Autophagy and genomic integrity. Cell Death Differ 20(11):1444–1454

Vijg J, Suh Y (2013) Genome instability and aging. Annu Rev Physiol 75:645–668

Wagster MV (2009) Cognitive aging research: an exciting time for a maturing field : a postscript to the special issue of neuropsychology review. Neuropsychol Rev 19(4):523–525

Whitmer RA, Sidney S, Selby J, Johnston SC, Yaffe K (2005) Midlife cardiovascular risk factors and risk of dementia in late life. Neurology 64:277–281

Zhu Y, Carvey PM, Ling Z (2006) Age-related changes in glutathione and glutathione-related enzymes in rat brain. Brain Res 1090:35–44